Daoism Excavated

Cosmos and Humanity in Early Manuscripts

WANG Zhongjiang

Contemporary Chinese Scholarship in Daoist Studies

Center for Daoist Studies, Peking University
北京大学道家研究中心

Three Pines Press
P.O. Box 530416
St. Petersburg, FL 33747
www.threepinespress.com

Chutu wenxian yu Daojia yanjiu de xin shiye 出土文献与道家研究的新视野. Translated by Livia Kohn.

Funding for this translation was provided by Mr. Kong Xiangping 孔祥平 of Yuanlianhe Real Estate. Management and interface coordination was supplied by Ms. Wang Lina 王丽娜.

9   8 7 6 5 4 3 2 1

First Edition, 2015
Printed in the United States of America
This edition is printed on acid-free paper that meets
the American National Standard Institute Z39.48 Standard.
Distributed in the United States by Three Pines Press.

Cover Art: Chinese traditional landscape. By Li Huanhuan and Brent Cochran.

Library of Congress Cataloging-in-Publication Data

Wang, Zhongjiang.
[Chu tu wen xian yu Dao jia yan jiu de xin shi ye. English]
Daoism excavated : cosmos and humanity in early manuscripts /
Wang Zhongjiang. — First edition.
pages cm. — (Contemporary Chinese scholarship in Daoist studies)
Includes bibliographical references and index.
ISBN 978-1-931483-62-9 (acid-free paper)
1. Taoism—Sacred books. 2. Taoist literature, Chinese—History and criticism. 3. Taoism—China—History—To 221 B.C. 4. Philosophy of nature--China--History—To 221 B.C. 5. Cosmology—History—To 1500. 6. Laozi. 7. Manuscripts, Chinese—China—Jingmen Shi. I. Title.
BL1900.A2W3813 2015
299.5'1482--dc23
2015002168

Contents

Preface

Vibrant scholarly exploration means dealing with unceasing change and opening to ever new visions. The field of Daoist studies is no exception. I personally have been deeply immersed in the exciting changes in this field, received much pleasure from its often unexpected new turns, and intensely felt the wonders of newly arising knowledge. In the 20th century, and particularly since the 1990s, China has seen the excavation of numerous manuscripts in bamboo and silk, including also many important documents related to early Daoism. This has offered amazing opportunities for the advancement of Daoist studies—opportunities I found too enticing to resist.

My academic career as a graduate student at Peking University began with research in modern Chinese philosophy, examining especially the thought of figures such as Yan Fu 严复, Jin Qiulin 金岳霖, and Yin Haiguang 殷海光. My focus was to gain a deeper understanding of how evolution theory and philosophical methodology developed in modern China. In this process, I came to appreciate how traditional Daoism approached issues faced by Chinese thinkers today. In order to provide a thorough explanation of just how Daoism transformed in modern China, I needed to dig much more deeply into its history, which made me study its early sources and, in due course, led me to its original thinkers.

Once I began to explore the transmitted literature of early Daoism, however, I found myself surrounded by a dense fog and beset by doubts. I feel very fortunate, indeed, that just around that time the bamboo and silk manuscripts became available. They not only allowed but even challenged me to engage in a complete rethinking of early Daoist thought and cosmology, offering a golden opportunity to resolve many difficult issues in its understanding. The manuscripts made it abundantly clear that there is so much more to early Daoism than previously thought and that established views in many respects have been inaccurate and even wrong.

This book presents the results of my inquiries toward a new and expanded understanding of early Daoist thought. Its title, *Daoism Excavated: Cosmos and Humanity in Early Manuscripts*, implies two things: first, that I take full advantage of the newly excavated bamboo and silk manuscripts; and second, that I present new, often radically transformed views on cosmos and humanity, in the hope to bring about a deeper understanding and a renewed appreciation of the ancient Daoist world.

I am deeply grateful to the National Social Science Foundation of China for its continued funding of my research into the newly excavated Daoist manuscripts. I also greatly appreciate the comments and suggestions I received from friends and colleagues as I presented my work at various international conferences. Thus, for example, I gave a talk on the *Fanwu liuxing*, which here appears in chapter 3, at a meeting of the Japanese Society for the Study of Excavated Materials and Chinese Culture in Tokyo, July 2010. I presented my findings on "The Han *Laozi* and Daoist Thought," here contained in chapter 7, at an international symposium held at Peking University, October 2013.

Most of the work contained in the other chapters, moreover, has appeared previously in various academic journals and edited volumes, whose editors often provided valuable feedback and stimulated new insights. More specifically, the introduction on Daoist cosmology was first published in *Zhongguo shehui kexue* 中国社会科学 (2013/5). Chapter 1 on cosmic emergence and unfolding in the *Hengxian* appeared originally in *Wenshizhe* 文史哲 (2008/2), while chapter 3 on the philosophy of the *Fanwu liuxing* was translated into Japanese—凡物流形』における「一」の思想構造とその位置—and published in the conference volume, *Shutsudo sairyō to kanji bunkagen* 出土資料と漢字文化圈(Tokyo: Jiko shoin, 2011).

Chapter 4 on the understanding of holding on to oneness in the *Huangdi sijing* was first appeared in *Huangdi sixian yu dao, li, fa yanjiu* 黄帝思想与道、理、法研究(Beijing: Kexue wenxian, 2013). Chapter 5 on the *Daode jing* line, "Dao models itself on self-being," was published originally in *Zhexue yanjiu* 哲学研究 (2010/8), while chapter 6 on the line, "A great vessel is late to complete," was first printed in *Jianbo wenming yu gudai sixiang shijie* 简帛文明与古代思想世界(Beijing: Beijing daxue, 2011). Chapter 7 on variant characters and new ways of reading Daoist thought as apparent in the Han *Laozi*, finally, came out in *Hubei daxue xuebao* 湖北大学学报 (2014/1). Only chapter 2, on the textual structure and conceptual layers of the *Taiyi shengshui*, was written specifically for this volume.

The presentation of all this work in English would not have been possible without the unstinting effort of many supporters. Among them, first of all, I would like to express my devout gratitude to Professor Chen Guying for his unfailing support and extensive help over many years of scholarship as well as his untiring efforts to make this translation series possible. Next, I wish particularly to thank Professor Livia Kohn, the translator of the book, who has undertaken this tremendous task with cheerful persistence, spending much time and effort in the process and never failing to impress me with her strong energy and vibrant aliveness. In a rare opportunity, I got to meet her in person in September 2014 in Luoyang, at the international symposium on "Laozi's Thought and Spirit: In History and the Contemporary World," an occasion I greatly enjoyed.

Professor Robin R. Wang, moreover, stands out as co-editor of the series and its ever-present caretaker. I am deeply indebted to her and wish to acknowledge her selfless dedication to the field and her unceasing work, without which the translation and publication of this book would not have been possible. Last but not least, I would like to thank Li Wei of Sun Yat-Sen University as well as the dedicated post-doctoral fellows at Peking University, Chen Zhibin, Pi Mimi, and Ai Chenyi, who helped with double-checking the English translation. It is because of their work and dedication that my research is now offered to the English-speaking world. I much wish that they share my joy and delight.

Wang Zhongjiang
Beijing, January 2015

Introduction

Daoist Cosmology in the Light of Excavated Manuscripts

A major philosophical breakthrough occurred between the traditional religion of the Three Dynasties (Xia, Shang, and Zhou) and the new worldview of the Eastern Zhou. It was nothing short of revolutionary. Closely related to the political and economic changes that accompanied the transition from a central government to a multiplicity of small states, it moved in two directions. For one, it shifted from mythical religion to the humanistic rationality of Confucianism; for another, it unfolded into the nature philosophy of Daoism.[1] Unlike in ancient Greece, where the philosophical breakthrough of the Axial Age went first to nature philosophy and from there to individual self-awareness (Jaspers 1953; 1989, 7-29; Xu 2006, 168-213; Yu 2005, 88-244), in China the two happened almost at the same time and gave rise to vigorous philosophical debates and the pluralistic competition of ideas.

There is little doubt about how Confucian humanism came about—notably in the light of the Guodian finds and matching materials in early ritual sources—but many questions remain about the development and unfolding of Daoism, not least the dating of its major early works. For example, quite a few scholars in both China and the West believe that the *Laozi* is actually later than the *Zhuangzi* 庄子 and did not emerge as a full text until the Qin and Han dynasties. This means that Chinese cosmology was formulated rather late, finding its first full expression in the *Huainanzi* 淮南子 of the 2nd century BCE (Ikeda 2009, 336-38; Graham 1989; 2003, 371-72).

Historical studies often give high priority to newly discovered materials, rightfully so, since it is often hard to come to new understandings based on transmitted sources alone. In addition, newly found sources invite new methodology and encourage alternate perspectives. The bamboo slips and silk manuscripts recently unearthed in central China are a case in point. This

[1] The Daoist concept of nature encompasses the two dimensions of a transcendent force and origin of the world and an inherent set of measures and standards to return to in life. While one is more pronounced in the *Laozi* and the latter is key in the *Zhuangzi*, I use the term "nature" to refer to both, including its full metaphysical understanding, inherent qualities, and existential relevance.

book, then, explores the *Laozi* 老子 and *Taiyi shengshui* 太一生水 from Guodian 郭店, the *Hengxian* 恒先 and *Fanwu liuxing* 凡物流形 from Shangbo 上博, as well as the *Huangdi sijing* 黄帝四经 from Mawangdui 马王堆.[2] These various manuscripts and fragments, used in conjunction with transmitted sources, offer new insights into the birth and unfolding of traditional Chinese cosmology and concepts of nature. In fact, their holistic discussion is becoming increasingly urgent, especially since previous studies were limited to certain kinds of texts.

Dating and Content

As much as these texts form part of the Daoist tradition, their expressed worldview is also representative of it. As much as they emerged during the Eastern Zhou, their cosmology reflects major beliefs of this time. A key question in this context, then, is where exactly to situate the *Laozi* and whether or not it is the first and original formulation of nature philosophy in China.

The discovery of its bamboo fragments at Guodian mark a major turning point. Much has already been written about its date. Based on archaeological studies, the Guodian tomb dates to around 300 BCE, that is, from the latter half of the Warring States. Two observations are obvious: first, the Guodian manuscripts cannot be any later than the tomb closure; and second, they must be earlier. In addition, since the manuscripts were hand-copied and some of the materials exist in other variants, they must go back to a longer tradition. This means that the text of the *Laozi* may well go back to the late Spring and Autumn period.

As Guo Yi, moreover, has already shown, the Guodian *Laozi* is the ancestor of all later versions discovered to date (1998). However, there are major discrepancies in the A and C versions (notably of ch. 64), which make it clear that the text existed in different strands and may even have had various original versions. This, in turn, means that the Guodian *Laozi* may be just one among several early redactions; it definitely shows that there were multiple strands and lineages of the work.

Now, the Guodian *Laozi* consists only of about 1,600 words, about one third of all other, later versions—the Mawangdui silk manuscripts, the Pe-

---

[2] Translator's Note: English translations of these texts are found in Henricks 2000; Cook 2012; and http://terebess.hu/english/tao/taiyi.html for the Guodian texts; Brindley et al. 2013 for the *Hengxian*; as well as Chang and Feng 1998; Ryden 1997; Yates 1997; and Lévi 2011 for the *Huangdi sijing*.

king University Han bamboo text, and the transmitted edition. Most scholars, in an attempt to explain this discrepancy, regard the Guodian *Laozi* as a portion of the other versions, but it remains unclear whether the other parts were supplemented later or just not contained in the tomb, nor do we know whether they were by the same or different author(s).

An important piece of evidence comes from a citation of the *Laozi* by Shu Xiang 叔向, which does not appear in the Guodian text. According to the "Jingshen" 敬慎 chapter in Liu Xiang's 刘向 *Shuoyuan* 说苑, Prince Ping of Han asks Shu Xiang a question about the respective merit of "hardness and softness." The latter replies with a reference to Lao Dan 老聃 and two citations from the *Laozi*: "The softest thing in the world overcomes the hardest" (ch. 43), and "People at birth are soft and weak; at death, hard and stiff" (ch. 76). Prince Ping is Han Xu 韩须, the son of Prince Xuan, a vassal of Duke Ding of Jin who ruled from 511 to 475 BCE (Xiang 1987, 245).

Shu Xiang, too was of the late Spring and Autumn period, a retainer of Duke Ping of Jin (r. 557-532). At the time of the recorded dialogue, Shu Xiang was already well over eighty years old. A contemporary of Confucius (Kongzi 孔子), he cites the *Laozi* with a passage that otherwise only appears in late Warring States materials, which means that the text was already known at that time and that passages beyond the 1,600 words found at Guodian go further back in history. Quite possibly, the *Laozi* already had about 5,000 words, and later additions were rather limited (see also Chen 2015).

The passage also gives testimony to the fact that Lao Dan and Laozi are one and the same person. This is also borne out in the *Zhuangzi*, which cites the two interchangeably—as does the *Hanfeizi* 韩非子, which incidentally is the first source to cite "Laozi" as a text rather than as a person. The yet unpublished Han *Laozi* at Peking University does the same and is also the first source to speak of a First and Second Book (*shang/xia jing* 上/下经) in this context. This shows that Laozi the person and *Laozi* the text form part of the same strand and tradition. To sum up, the various excavated materials present us with four distinct *Laozi* texts, showing just how Laozi's thought unfolded from the Eastern Zhou to the Qin and Han.

Besides the *Laozi*, excavated manuscripts also include three early texts and one set of later materials. The first early text is the *Taiyi shengshui*, discovered at Guodian in the same cluster of bamboo slips as the C version of the *Laozi*. It thus dates from the same period, and most likely goes back to Laozi's disciple Guanyinzi 关尹子 (aka Yin Xi 尹喜) (see Li 1998). The text clearly shows influence of the *Laozi* in both its key concepts, "one" and "water," which it expands to indicate the underlying "great oneness" and core creative factor of the universe. Its emphasis on the notion that "the Dao of heaven values weakness," too, connects to Laozi's favoring weakness and softness over aggression and competition. The Guodian *Laozi* even has a section heading, "Weakness is the Function of Dao" (ch. 40 today).

The two other early texts are the *Hengxian* and *Fanwu liuxing* from Shangbo, the tomb of a Chu noble. Archaeological carbon dating places this find to about the same period as Guodian, i.e., 2257 ± 65 years before present (Zhu 2002, 3; Ma 2001a, 2). The three texts are thus part of the same tradition, placed as they are "between Laozi and Zhuangzi" (Li 2006, 494-95).

The later document is the *Huangdi sijing.* A collection of four treatises from the Huang-Lao 黄老 school of the late Warring States period (Tang 1975), it consists of *Jingfa* 经法 (9 chs.), *Shida jing* 十大经 (15 chs.), *Cheng* 称 (1 ch.), and *Daoyuan* 道原 (1 ch.). Overall, it presents a synthesis of Daoist thought with other major philosophies of the time, notably Legalism, Confucianism, and Mohism. Like the *Zhuangzi*, *Liezi* 列子, *Guanzi* 管子, and *Heguanzi* 鹖冠子 (Dai 2000, 15-39), it represents a later stage of Daoist thought.

Last but not least, there are the two *Hanfeizi* chapters, "Jie Lao" 解老 and "Yu Lao" 喻老 (chs. 20-21), so that the complete list of early Daoist cosmological materials is as follows: *Laozi* → *Taiyi shengshui* → *Fanwu liuxing* → *Zhuangzi* → *Guanzi* → *Huangdi sijing* (*Daoyuan*) → *Heguanzi* → *Hanfeizi*. All these texts go back to the Eastern Zhou and accordingly reflect the philosophy and cosmology of this period, clarifying just how revolutionary the change was from the mythical religion of the Three Dynasties.

Generally speaking, myth and religion, science and philosophy all provide explanations on how the world came into being, but the general tendency is that the earlier the period in human history, the more dominant were myth and religion and the weaker were science and philosophy. A crucial change occurred around 600 BCE, in what Karl Jaspers has called the Axial Age (1953). In China, too, at this time, the myth-based, religious cosmology of the Shang and Western Zhou had weakened considerably while philosophy began to rise. Although history is not generally linear or follows a single stand, this overall tendency is clearly visible in both East and West.

Among the various philosophers of the Eastern Zhou, moreover, Daoists had the strongest predilection and richest interest in questions regarding the origin of the cosmos and all beings. They were the most enthusiastic in pursuing heavenly writings and proposing solutions to cosmic riddles. For this reason, they came to lay the foundation of nature philosophy in China, influencing even the *Xici* 系辞 commentary to the *Yijing* 易经 as well as the *Xunzi* 荀子.

Laozi and other early Daoists replaced the major Three Dynasties cosmic powers of the thearch (*di* 帝) and heaven (*tian* 天) with Dao and oneness, which their successors turned into the foundation stones of Chinese cosmology. The Daoist tradition is thus at the center of the development of early Chinese nature cosmology. As clearly expressed in *Laozi* 25, they defined Dao as the underlying force of the universe. Under its influence, the three early manuscripts developed the concepts of oneness (*yi* 一), Great Oneness (*taiyi* 太一), and Constancy Before (*hengxian* 恒先) in an effort to describe the origin of the myriad beings. The *Daoyuan*, too, uses Dao to speak of the root of all beings and identifies it with oneness. The *Zhuangzi*, in addition, in its development of Laozi's Dao, emphasizes the notion of the "transformations of energy (*qi* 气)," while the *Guanzi* provides explanations of both Dao and virtue, and the *Hanfeizi* combines Dao with principle. All these are ways in which early Daoist cosmology developed during the Eastern Zhou. Let us now look at these features in some detail.

Cosmic Unfolding

Examining the various explanations offered by philosophy and science for the origin of the universe, it becomes obvious that scientific theories today are gaining in importance, while philosophical concepts are increasingly less relevant. In antiquity, the opposite was true, and philosophical theories were more widespread. Even more amazingly, the cosmologies of contemporary science and of ancient Chinese philosophy bear some strong similarities. As Dong Guangbi outlines, many new Daoist thinkers today come from the natural sciences, following in the footsteps of Fritjof Capra and linking modern physics with Oriental mysticism (see Dong 1991; Capra 1975; 1999). Their premises are obviously vastly different and we have to be careful with one-on-one comparisons. Yet, to say that they have nothing in common would also be very shortsighted.

Ancient Daoist thinkers firmly believed that the universe and all beings had a particular source or origin, which they called Dao, oneness, Great Oneness, Constancy Before, etc. This cosmic origin to them was also the mother of all existence. While they insisted on a specific source, however, they also emphasized the enormous dimensions and infinity of the world, so that both the *Zhuangzi* and the *Mozi* 墨子 speak of space and time as limitless.

Philosophers generally distinguish concepts according to time and logic. Cosmogony clearly belongs into the first category, as something that occurred earlier in time. Both ancient Daoists and contemporary scientists posit that the universe has a beginning or primordial state, but while scientists see this as a singular point, Daoists understand it to be a multilayered phenomenon. Depending on which aspect of this they emphasize, moreover, they describe the inherent nature and unfolding of the universe differently.

The excavated texts present four distinct models. First, the *Fanwu liuxing* describes universal unfolding in terms of beginning with oneness, then moving to the two, the three, the mother, and eventually resulting in "combination" (*jie* 结). The *Hengxian* speaks of Constancy Before, from which things evolve through space (*huo* 或), energy (*qi* 气), material existence (*you* 有), and beginning (*shi* 始) to the passage of time (*wang* 往). The *Taiyi shengshui* begins with Great Oneness, which gives birth to water, from which arise heaven and earth, which in turn connect back to Great Oneness and water. Once established, spirit and light emerge, in due course giving rise to yin and yang, the four seasons, cold and heat, dampness and dryness, to result in the ongoing cycle of the years (*sui* 岁). The *Huangdi sijing*, finally, has the universe begin with oneness, from which spring heaven and earth, yin and yang, the four seasons, and hard and soft, to result eventually in the myriad beings.

Transmitted sources add their own versions. Thus, the *Laozi* has the well-known sequence from Dao to oneness and on to two, three, and all beings (ch. 42). The *Wenzi* 文子, not unlike the *Huangdi sijing*, sees life arising from oneness, then move on to heaven and earth, the four seasons, yin and yang, humanity alone, and hard and soft, to eventually result in the myriad beings (ch. 3, "Jiushou" 九守). The *Heguanzi*, finally, starts it with the one energy and moves through intention, planning, naming, form, affairs, connections, and seasons to beings (ch. 5, "Huanliu" 环流). All these different models can be understood and analyzed from a variety of different perspectives.

Let us first look at what they propose in terms of origin. They speak variously of Dao, oneness, Great Oneness, constancy, and so on, but why do they use so many different terms, and what exactly is the relationship between them? The three early manuscripts do not speak of Dao at all, but refer to oneness or constancy, possibly to avoid being lumped together too closely with the *Laozi*. Thus, the *Hengxian* only uses Dao only once, in the combination "Dao of heaven," while the *Fanwu liuxing* speaks of it variously, but consistently places greater emphasis on oneness. The same also holds true for other, traditionally transmitted sources.

Still, the various terms early Daoists use for the beginning of the universe are quite compatible with each other. Laozi uses primarily Dao, but he also employs oneness, as in the expression "oneness of heaven." Following in his wake, Daoists honor Dao as their key concept, yet they also elevate oneness to the notion of Great Oneness. Examples include the *Daoyuan* of the *Huangdi sijing* with its statement that "one is the core number," and the "Chengfa" 成法 chapter (ch. 9) of the *Shida jing* in the same collection with its claim that "oneness is the root of Dao."

Oneness also plays a role in the *Fanwu liuxing*, while the *Taiyi shengshui* elevates it to Great Oneness. According to one interpretation, the latter refers to the star god of the North Culmen (Beiji 北极) (Li 1999, 297-300), but that link came after its importance as a philosophical concept and was a secondary modification (see Qian 1932). Transmitted sources, too, for the most part accept the notion that both oneness and Great Oneness refer to Dao. Another term for the universal origin is Constancy Before, the title of the *Hengxian*. Its close relation to Dao is apparent in the *Daoyuan*, which explains one with the other. Its primary goal is to "inquire into Dao," but when it speaks of Dao's beginnings, it uses Constancy Before. Both "constancy" and "before" are terms related to time. However, the latter is not used in its ordinary sense as coming before something else, but rather in a more cosmic manner as the state of the universe before the beginning of all.

Whichever term they use, Daoists in all cases indicate the root and original state of the universe. Although at first glance quite different from dominant Western notions and descriptions of the beginnings of humanity, there is quite a bit of similarity with ancient Greek philosophies of nature, although the latter have a great deal less interest in cosmological speculation. Still despite the large variety of different terms and concepts, in many ways, both would agree to Laozi's formulation:

> There is a state, in chaos yet complete;
> It comes first, before heaven and earth.
> Silent it is, and solitary;
> Standing alone, it never changes.
> We can consider it the mother of all under heaven.
> I do not know its name.
> To call it something, I speak of Dao. (ch. 25; Guodian A; see Henricks 2000)

Here the term "Dao" indicates the initial state of the universe. The transmitted version of the *Laozi* further notes that Dao cannot be described (ch. 1), although it "may appear as a being, yet it is vague and obscure (ch. 21). It is "simple and nameless" (ch. 32), "hidden and ineffable" (ch. 41), and so on. All these are expressions for the original state of the universe, Dao at the beginning of creation.

Both the *Hengxian* and the *Huangdi sijing* develop Laozi's basic notions. The *Hengxian* begins by speaking of the state before being (*wuyou* 无有), describing it in terms of simplicity (*pu* 朴), stillness (*jing* 静), and emptiness (*xu* 虚), emphasizing each by adding the epithet "great" (*tai* 太), meaning to the highest degree, ultimate, utmost. It says, "Before there were heaven and earth, before there was any doing, moving, emerging, becoming, emptiness and stillness rested in oneness, serene and peaceful, still and unified, without any light, without any arising." Based on this, it becomes clear that the original state of the universe was long-lasting, "a constant and permanent state." The *Daoyuan* echoes this,

> At the initiation of constant nonbeing,
> All was merged in great emptiness.
> Empty and same, all was in oneness,
> Constant oneness, nothing more.
> Misty and blurred,
> There were neither light nor darkness.
> Spiritual and subtle, yet it filled everywhere;
> Quintessentially still, it was not luminous. (see Yates 1997, 173) [3]

This strongly emphasizes the state of "great emptiness" at the beginning of creation, how all was merged and undivided, without light, in total oneness. Daoist descriptions of the original state thus do not make use of concrete terms but speak of it as being without name, without form, without image, greatly empty, merged in chaos. Yet the state also contains the potentiality of all and is the greatest font of all existence.

Still, we have to take this a step further and realize that the original creative state of the universe is real, it is something: Dao, oneness, Great Oneness, or Constancy Before. These are highly abstract concepts, despite the fact that they are also expressed in metaphors such as the "mysterious female" or the "spirit of the valley." Daoism does not have anything like the ancient Greek concept of the elements, like water or fire, nor does it have ideas of things like metal and earth, later part of the five phases. Naturally, it also lacks any version of the Greek notion of "atoms" or other solid building blocks.

Although Daoist thinkers often use the negative "non" to describe the original nature of the pre-creation state, yet—as is well known—they do not mean "pure empty nonbeing" but rather "ultimate being." This is evident in

---

[3] Translator's note: The word "see" in reference to previous translations of ancient sources indicates that this work was consulted, i. e., used with adjustments. For the *Laozi*, I have consulted Addiss and Lombardo 1993; Chan 1963; and Legge 1962.

Laozi's words, "The myriad beings carry yin and embrace yang;" in Zhuangzi's statement, "The one energy pervades all under heaven;" in Guanzi's notion of "essential energy;" in Liezi's account that "Great Antecedence is the beginning of energy;" as well as in Han visions of primordial *qi*—expressed as "constant energy" in the *Hengxian*. It is thus best to think of "ultimate being" in terms of cosmic life energy—despite the fact that the sources do not speak in these terms to begin with.[4]

Thus, different generational models speak differently about the mode of universal becoming. Some being more abstract than others, it is often hard to determine what they mean exactly, particularly the *Laozi*, *Fanwu liuxing*, and *Hengxian*. Some are so abstract, moving forever from complexity to intricacy, that they cannot be understood by themselves, but require recourse to other sources. For example, in the classical passage of the *Laozi*, "Dao gives birth to one, one gives birth to two, two give birth to three, and three give birth to the myriad beings" (ch. 42), what exactly and in concrete terms is meant by going from two to three? The *Huainanzi* provides the earliest interpretation, "Dao begins with oneness. Oneness alone, however, does not give birth. Therefore, it divided into yin and yang. From the harmonious union of yin and yang, the myriad beings were given birth" (3.28; Major et al. 2010, 133). That is to say, in this reading. "two" is an expression for yin and yang, while "three" indicates their joining.

The *Hengxian* model is different. "Space arises. Once there is space, there is energy. Once there is energy, there is material existence. Once there is material existence, there is a beginning. Once there is a beginning, there is the passage of time." None of these, with the exception of energy (*qi*), is at all clear in concrete terms—they are all highly abstract, especially "space" (*huo*) and "material existence" (*you*). It is possible to read "space" as the open state between Dao and energy (Li 2003, 288) or read it as "domain" (*yu* 域), which would indicate the cosmos (*yu* 宇) (Li 2003, 81-82; Wang 2008). But the later state of "material existence" remains utterly unclear. I am inclined to read it in terms of a great state of initial form, i.e., an expression for heaven and earth. "Beginning" and "the passage of time," then, might refer to the myriad beings as birthed by heaven and earth plus their transformations and unfolding.

The *Fanwu liuxing* echoes the *Laozi*, but phrases it in three-character lines and adds a phase of "mother" before its final state of "combination." All

---

[2] On Greek concepts of nature, see Collingwood 1945; 1999, 31-99. For ancient Indian notions of creation through water, see Wu 2000,103-05, 159-62.

terms it uses are highly abstract, and despite its similarities to the *Laozi*, it really presents its own vision, which may or may not be reconcilable with ideas of cosmic energy. As different as these various models are, a key way of understanding them is by comparing the various concrete terms and metaphors they employ in their description of the gradual unfolding of existence from one level to the next.

Important supplementary information comes from the other manuscripts and some transmitted sources, including the *Taiyi shengshui*, the *Huangdi sijing*, and the *Wenzi*. They all tend to describe the universe as moving from Great Oneness to somewhat more concrete entities, notably water, heaven, and earth. In all cases, the models give evidence to the urge of the ancients to identify a pivotal factor and systematic evolution of nature.

All early cosmologists believed that the universe took its beginning from a foundational natural entity and created life in a particular order, but they had a hard time determining what exactly that central pivot or foundation was. In addition, while they all insisted that simpler factors generated more complex structures (e.g., wood being shaped into different types of furniture), the more concrete entity they proposed at the origin, the harder it was for them to see it evolve into complexity (e.g., once wood is shaped into a chair, it is hard to transform into a table).

Be that as it may, at this point we can distinguish two dominant tendencies: to see cosmic unfolding through a figurative versus abstract model. Within this framework, Daoist cosmology proposes that the universe underwent a step-by-step evolution, transformation, or emergence. This is cosmogony, distinctly different from visions of active creation or intentional bringing forth. The Daoist vision of universal emergence and cosmic structure always moves from darkness to light, from chaos to order, from simplicity to complexity, from the one to the many. Within this, it proposes various abstract or figurative (concrete) factors to outline levels and goals, increasingly differentiating beings from chaos in the process of arising.

The *Laozi* and the *Fanwu liuxing* both use the concept of "giving birth" (*sheng* 生) to describe the arising of the universe. The *Huangdi sijing* first uses "becoming" (*cheng* 成), and only then speaks of "giving birth," using the term to refer to a higher and more complex level in the process. The original meaning of "giving birth" goes back to plants sprouting from the soil; from there, the word expanded to include the raising of livestock and children. Based on the natural birthing process of human life and natural phenomena, Daoists extrapolated the unfolding of the universe in terms of being birthed by a "mother." Compared to the contemporary vision of universal beginnings in terms of the Big Bang, the vision of giving birth is warmer and more integrative, laying the foundation of an organic understanding of the world.

The word *cheng*, then, has implications of change, transformation, unfolding, and completion. Compared to the rather active, even forceful "giving birth," it expresses more the silent and gentle transfiguration of beings and situations. Most texts, other than the *Hengxian*, tend to speak of "once there is. . . ." Words like "moving" and "being" imply a sense of emergence and birth, arising and transforming. Since Daoists never suggested the existence of an active creator, of a force beyond nature that would produce things, the origin and arising of the universe as well as of all beings must rest in itself, go back to its own inherent way (Dao) or underlying unity (Great Oneness). Thus, the *Hengxian* says, "It fulfills itself without repressing itself," and the *Zhuangzi* speaks of universal "self-stimulation," "self-creation," and "self-transformation."

Virtue and Differentiation

The Daoist vision of creation is essentially cosmogony, but it is also a form of ontology or, as Zhang Dainian calls it, a theory of foundations (1982, 6-24; 2002). The two closely merge in Daoism, which makes it significantly different from the philosophies of Plato and Aristotle as well as harder to define as a single system. Multiple versions of cosmogony preclude a systematic ontology founded on logic and reason that serves to explain the ultimate cause and inherent nature of all beings.

A key concept in the Daoist context, moreover, is virtue or inner power (*de* 德), prominent in all its different strands and often identified with Dao. It plays a central role in the Daoist vision of living beings, since the relation between Dao and virtue serves as the model of that between the primordial origin and all life. Still, it only represents one aspect of the feature, others including the relation between Dao and beings, simplicity and complexity, and the one and the many. The various excavated materials are essential in providing a fuller understanding of these dimensions.

Unlike Confucian thinkers who focus centrally on the human mind, Daoists are primarily interested in the original nature of living beings and develop their general theories from here. How, then, can this original nature be best understood? Scholars often explain it in terms of virtue, inherent in all beings and affairs, a form of secondary application or concrete manifestation of Dao—but they never quite say how Dao comes to be so (Chen 1992, 13-14; Luo 2005, 91-92).

Contemporary cosmologists claim that nature and living beings evolved gradually over a vast time span. Their original nature accordingly forms part of this evolution. It is open to discussion just how much evolution there is in ancient Daoism, despite the fact that Zhuangzi clearly speaks of the transformation of energy, giving quite a few concrete examples, and Hu Shi thinks of him as an early representative of progress thinking. Still, Daoists in general are not that interested in understanding just how each being came to be what it is and how diversity arose on the planet.

To them, the nub of the matter lies in the relationship of the originally unified Dao, simplicity, or oneness to intricate virtue, complexity, and multiplicity—the former pointing to the inherent original nature of all things while the latter express their concrete manifestation (Ikeda 2006, 15-30). Thus, the *Laozi* says, "When uncarved wood breaks up, it forms vessels" (ch. 28); "Dao in the world can be compared to rivers and streams flowing into the sea" (ch. 32); and "Great Dao flows everywhere; it may go left or right" (ch. 34). Similarly, Zhuangzi notes, "Dao does not falter before the huge, is not forgetful of the tiny; therefore, the myriad beings are complete in it" (ch. 13) and "Dao: there's nowhere it is not" (ch. 22). Passages like these show that Daoists think of the different qualities and manifestations of things as the result of Dao "breaking up" or "dividing."

The *Laozi* does not yet make the common later connection of virtue and its homophone *de* 得, "to attain." Here, "attain" only means that individual beings realize certain particular characteristics and traits on the basis of all-pervasive oneness. The classical passage has, "heaven attained oneness and became clear; earth attained oneness and became tranquil" (ch. 39). The general understanding is that "oneness" means Dao and these various entities realize Dao. Since beings are different, they inevitably gain different things when attaining oneness. In other words, the passage speaks of how different beings attain different part of universal oneness and how they relate to Dao differently. It reflects but does not explain the inherent interconnectedness and fundamental multiplicity of all that is.

Like the *Laozi*, the *Fanwu liuxing* says that "there is oneness" and notes that beings "attain it" when it speaks of the difference among living beings, but it still does not connect "attain" with virtue like the *Daoyuan* of the *Huangdi sijing*. It examines the manifold things of existence—sun and moon, thunder and lightning, wind and rain in heaven as well as water and fire, grasses and trees, birds and beasts, soil and humans, ghosts and spirits on earth—and asks over forty questions about them. For example,

> All beings flow into form, but how do they attain it? They flow into form and become bodies, but how can they avoid dying? Humans flow into form, but

> how do they attain it? They flow into form and become bodies, but how do they lose them again and die? How do grasses and trees come to grow? How do birds and beast come to sing?

While we can offer scientific explanations to these questions today, the text starts out with the basic assumption that there is some common ground and that all beings somehow attain it. For example, when it asks why there are different species in the universe and the world, it says, "There is oneness; if the world perishes, it no longer exists. If oneness perishes, the world also loses its existence." Similarly, when it asks why plants grow and animals make sounds, it says that "plants attained it and came to grow; animals attained it and came to make sounds." In other words, all beings came to be what they are because they "attained" the function of oneness and Dao.

In contrast to this, the *Daoyuan* has no interest in how Dao relates to the concrete factors of the world, but instead focuses on the broader connection between Dao, living beings, and affairs. Still, it works with "attain":

> It takes oneness as its measure and never changes,
> And is right there even for crawling insects.
> Birds attain it and fly.
> Fish attain it and swim.
> Wild animals attain it and run.
> The myriad beings attain it and live.
> The hundred affairs attain it and are complete. (see Yates 1997, 173)

This is also reflected in the transmitted sources. For example, the *Zhuangzi* says, "Dao has its reality and its signs . . . Xiwei attained it and held up heaven and earth" (ch. 6). This shows that the core idea of Dao is precisely that it is immanent in all beings, endowing them each with unique characteristics and abilities. While the *Zhuangzi* points to humanity as playing a prominent part in the greater picture, it yet insists that all beings attain their particular nature from Dao.

Another transmitted text that echoes this is the *Hanfeizi* in its discussion of the relation of Dao to beings and principle (*li* 理). Here Dao is the root cause "that makes the myriad beings just so," and "causes them to grow." It is their "specific inherent principle" that makes them unique. The text lists a number of different things as examples to prove the point. "Heaven attained it to be high; earth attained it to be containing" (ch. 20). In both the use of "attain" and their overall vision, the two texts are very close.

In all these Daoist discussions of the inherent characteristics and shapes of living beings, "virtue" plays no role whatsoever. They do not oppose or connect Dao to it, but contrast Dao with beings, simplicity with complexity, and the one with the many. Nevertheless, there is a link, most clearly expressed in the *Laozi*: "Great virtue flows from Dao" (ch. 21); "Who follows Dao joins Dao, who follows virtue joins virtue" (ch. 23); "When Dao is lost, virtue arises" (ch. 38); "Dao gives birth to them, virtue raises them . . . ; therefore the myriad beings venerate Dao and honor virtue" (ch. 51). But while virtue is present in all beings here, there is no sense of any "attaining" taking place.

A typical expression how things attain virtue from Dao appears in the *Guanzi*: "Virtue is the lodge of Dao. All beings attain it to take birth . . . For this reason, 'virtue' is 'attain,' and 'attain' means 'to realize one's natural so-being'" ("Xinshu shang" 心術上). That is to say, "virtue" here not only means the characteristics and traits individual beings get from Dao, but it also takes on the role of the force that "nurtures" and "raises" them, echoing the *Laozi* (ch. 51).

Xu Fuguan interprets "raise" to mean "produce" (2001, 298), but that may not be entirely correct. To me, it is more like "support," as the word *chu* 畜 appears in the *Shijing* 诗经: "But you do not support me, and I go back to my country and clan" ("Xiaoya" 小雅). This is echoed in the *Laozi*, when it speaks of Great Dao as "clothing and feeding all beings without claiming to be master over them" (ch. 34). Here the meaning is clearly "nurture." The same tenet also appears in the *Guanzi*, which states, "What nurtures the myriad beings is called virtue" ("Xinshu shang"); and the *Wenzi*, "That what gives birth to beings is Dao; what grows them is virtue. It supports them and nurtures them, follows them and grows them, benefiting all without preference, in harmony with heaven and earth—this is what we call virtue" (ch.5, "Daode" 道德).

If we assume that Dao is the highest force of support, then virtue is its concrete aspect, a mothering, raising power essential for the life and growth of all beings. The *Zhuangzi* says, "That by which beings come to life is virtue. . . Form without Dao is not alive, life without virtue does not shine" (ch. 12); and, "Dao is the idol of virtue; virtue is the radiance of Dao" (ch. 23). The word *sheng* 生 in connection with virtue is different from when it appears linked with Dao: meaning "life" rather than "giving birth" and pointing to enhancement rather than existence. Virtue in this sense, then, is the power of protecting and enhancing the life of all things and beings, connecting closely to its understanding as supporting force. This, moreover, holds not only true for life within individual entities, but also applies to their interaction, so that each thing or being in the process of life rests in constant interchange with all others.

The Han dictionary *Shuowen* 说文, compiled by Xu Zhen 许慎 (ed. Beijing: Zhonghua 1983), explains "virtue" as "what one attains on the outside from other people and on the inside from oneself." Daoists, on the other hand, do not limit this feature to humans but expand it to all things and beings, both in terms of internal quality and external exchange. "Virtue" here comes to mean the inherent power that nurtures and supports all. Most fundamentally, Daoist thinkers take the unlimited nature of Dao, as well as its open flexibility, and use it to explain how oneness or unity at the highest level can divide into the multiplicity of things and manifest in their vastly different characteristics.

The *Huainanzi* provides a further clue to this understanding, claiming that beings emerge differently from oneness because they take form as different separate entities (*fenwu* 分物). "Together emerging from oneness, so that each acquired its distinctive qualities, there were birds, there were fish, there were animals: this we call the differentiation of things" (14.1; Major et al. 2010, 536). Neo-Confucians of the Song and Ming dynasties express the same idea when they speak of "the principle of oneness separating into differentiation" to explain the relationship between the one underlying principle and the multiplicity of existence in an expansion of the *Hanfeizi* notion of Dao as principle.

When Daoists relate Dao with beings, simplicity with complexity, and the one with the many, they not only look for particular laws and rules but also for the connection of unlimited potential and matter to the concrete abilities and physical nature of all beings. In this way, they can explain how the myriad beings attain their particular abilities (principle) as well as their unique shape (energy, matter) from Dao. To borrow Plato's concept of particularity, the myriad beings are not all the same because they cannot possibly share in the Dao in the same way. However, the particularity of separate entities here does not carry the Platonic implication of the world of forms or ideas that is beyond reality and devoid of solidity nor is it a mere concept without physical substance. It is both ideal and real, physical and metaphysical.

Unlike the Christian concept of God taking a particular embodiment in the world, Dao in China is spread throughout the myriad beings and shared among them. "Dao forms the body of all beings," they say, but again this is not the same as the notion of divinity spreading everywhere as formulated in Spinoza. For Daoists, Dao spreads throughout living beings—transcendent and immanent at the same time, theory and practice integrated.

Nonaction

How life attains its particular characteristics and how the myriad beings come to life are two different questions. In other words, after the myriad beings take birth, they develop in their unique way; yet before they take birth, they are already differentiated as specific entities. This is just like animals in nature: they undergo two different stages, before and after birth. In the context of the relationship of Dao and world, then, how exactly does Dao connect to, or contrast with, living beings and how does it impact their particular life? Daoists in this context propose the familiar notions of nonaction (*wuwei* 无为) and naturalness or "self-being" (*ziran* 自然), which have been much discussed in scholarly literature. Here, I would like to reexamine their relevance in connection with the concepts of weakness and spontaneous unfolding.

First of all, is nonaction in the *Laozi* an inherent characteristic of Dao or is it a mode of its functioning? The term appears most frequently in connection with the sage king and only twice related to Dao.

> Dao always rests in nonaction, and there is nothing that is not done. (ch. 37)

> Learning means to increase daily;
> Practicing Dao means to diminish daily.
> Diminish and again diminish: this leads to nonaction,
> In nonaction there is nothing that is not done. (ch. 48)

The first passage, in the Mawangdui manuscripts as well as in the Han *Laozi*, reads, "Dao is constant and without name." The second passage, in the manuscripts, does not have the characters for "nonaction." This not only means that the idea of "do nothing and nothing is not done" may well be a later addition (Gao 1996, 54-57), but also that nonaction may originally not be closely linked with Dao, rendering the entire idea of Dao as nonaction questionable.

On the other hand, the Guodian C version has the full phrase and matches the passage now in chapter 37. It says,

> Dao is constant nonaction.
> If lords and kings hold on to this,
> The myriad beings self-transform.
> As they transform, they want to act.
> At this point,
> Take care to maintain the simplicity of the nameless.

> Also, make sure to know when it is enough.
> Knowing when it is enough affords stillness.
> The myriad beings will be self-stabilizing.

This clearly sees nonaction as a function of Dao; it also identifies the two. The transmitted version in addition has the phrase, "and there is nothing that is not done." This means that Dao moves by not acting yet attains results in the most direct way.

The phrase also occurs in chapter 48 of the transmitted version, which is very close to Guodian C and was not added later:

> Learnying means to increase daily;
> Acting in Dao means to diminish daily.
> Diminish and again diminish
> Is being in nonaction,
> And in nonaction, there is nothing that is not done.

According to this, following Dao in one's actions means using nonaction, which leads to a state where nothing is not done. Nonaction here is clearly a fundamental function of Dao. The *Zhuangzi* cites the same passage, combining the manuscript and transmitted versions: "Acting in Dao means to diminish daily. Diminish and again diminish leads to nonaction, and in nonaction there is nothing that is not done" (ch. 22). Clearly, the last phrase is not a later addition but goes back to an ancient original.

Softness and Weakness

Another key feature of Laozi's thought, later picked up in the *Lüshi chunqiu* 吕氏春秋 ("Buer" 不二) is valuing softness. The *Laozi* tends to combine softness with weakness, emphasizing their practical application—in clarity and stillness, lack of competitiveness, namelessness, simplicity, and the like. These various concepts all express different angles of nonaction, providing models of how a Dao of constant virtue acts in relation to the universe and the myriad beings. Being soft and weak is essential: "Reversion is the movement of Dao; weakness is the function of Dao" (ch. 40). Guodian A has this as, "Reversion is Dao moving; weakness is Dao functioning," showing that Laozi deeply appreciates weakness as a particular ability and application of Dao.

Understanding nonaction from the perspective of softness and weakness, it becomes clear that the way Dao interacts with the myriad beings it has given birth to is by being gentle and flexible, light and subtle, never intentionally doing or undertaking anything. Undertaking things to a minimum is the opposite of forceful action, control, and dominance. The central point in the Daoist theory of existence, then, is that the root power at the core of life pervasively moves and functions in the softness and weakness of nonaction, allowing the myriad beings to fully realize their inherent nature and live to the utmost. Thereby it creates order and harmony in the cosmos.

Later Daoists expanded this theory. For example, the *Daoyuan* describes nonaction as the root of Dao and says,

> Oneness is its appellation;
> Emptiness is its dwelling.
> Nonaction is its element;
> Harmony is its function. (see Yates 1997, 175)

"Element" here is another expression for "root," as the *Huainanzi* says, "Nonaction is the substance of Dao" (14.24).

Similarly, the *Taiyi shengshui* says, "The Dao of heaven values weakness," a phrase not found in the other texts but clearly connected to Laozi's notion of valuing softness. The *Laozi* frequently emphasizes this feature, using terms like softness, harmony, female, flexible, and weakness, and opposing any form of force and rigidity. It is thus no wonder that the *Lüshi chunqiu* describes its thoughts as centering on "valuing softness," which echoes the overall Daoist favoring of nonaction as.

Following its statement that "the Dao of heaven values weakness," the *Taiyi shengshui* has, "Shaping things to form in order increase means fighting with force, compelling with . . ." Here seven characters are missing. Li Ling thinks they match the pattern of the phrase, "XX with YY." He says, "Fighting and compelling both use force. Increase is accumulation, but what really helps increase is weakness and humility. The text means, in fact, the opposite" (2002, 32-33). On this basis, we can supplement the missing characters: "Compelling with rigidity, XX with weakness, XX with softness." Both "fight" and "compel" are negative terms, while actions involving weakness and softness are positive—lending full expression to the notion that "the Dao of heaven values weakness."

Dao originally functions with nonaction, softness, and weakness in the universe to allow the myriad beings to develop and live fully in accordance with their inherent nature. According to the *Laozi*, Dao gave birth to the myriad beings and gave them both virtue and various abilities as part of their inherent nature, but it would not control, suppress, or push them. "Dao produces them but does not take possession of them. It acts on them but does not presume upon them. It leads them but does not master them" (ch. 51), passages emphasized already by Liang Qizhao (1999, 3116-19) and Bertrand Russell (1966; 1996, 146-54). The text also says,

> The myriad beings depend on Dao for life,
> But it does not turn away from them;
> It accomplishes its task but does not take possession.
> It clothes and nurtures the myriad beings,
> But it does not become their master—
> We call this small.
> The myriad beings return to it,
> But they do not honor it as their master—
> We call this great. (ch. 34)

In other words, to make sure the myriad beings can develop and live fully in accordance with their inherent virtue and abilities, Dao serves as the deepest root and highest unity of all beings. It is their protector and guide, always kind, always compassionate. "It skillfully provides for all and brings them to perfection" (ch. 41). It has an endless supply mysterious virtue, "being impartial, it is kingly" (ch. 16). It also keeps everything in balance, "diminishes excess and supplements deficiency" (ch. 77) and "without favorites rests in virtue alone" (ch. 79).

Natural Self-Being

How, then, do the myriad beings live and work? In a natural, self-generating, and self-organizing order.[5] By actively applying the nonaction, softness, and weakness of Dao, the model for life of the myriad beings is naturalness or self-being—their self-generating activities and states of life, what Westerners describe as "nature" (Lovejoy 1936; 1996, 567-80). When Laozi speaks of "Dao models itself by being just so" (ch. 25), he in fact says that "Dao follows the nature of the myriad beings," which ultimately means it lets all beings "inherently ollow what is right for them."

The idea of "nature" in this context is not a fixed concept, nor is the fundamental nature of Dao, since it essentially matches this "nature." Essentially counterintuitive, this idea challenges us to change our entire percep-

[5] This concept goes back to Hayek 1975; 1998. For applications to Daoist thought, see Zheng 1998, 236-98; Ikeda 2009, 527-98.

tion. Guodian C speaks of being able "to support the self-being of the myriad beings and never daring to act [intentionally]," a direct testimony to the understanding of self-being as the way of life of the myriad beings (see Wang 2010a). Guodian A further says,

> Dao is eternal and has no name.
> Although its simplicity seems insignificant,
> None in heaven and earth can master it.
> If lords and kings held on to this,
> The myriad beings would naturally submit to them.
> Heaven and earth unite to drip sweet dew.
> Without the command of human beings,
> It drips evenly everywhere. (ch. 32)

The idea that they "naturally submit" further connects to Guodian A's notion that they "self-transform" and "are stable in themselves." This allows us to take the argument one step further and read self-being as self-transformation and inherent stability, clarifying just how the myriad beings live and work in the world.

The "Daofa" 道法 chapter of the *Jingfa* contained in the *Huangdi sijing* further expands this by saying that "the myriad beings naturally have names" and "of themselves do what is right," enhancing the *Daoyuan's* notion that "all beings are stable in themselves" in describing the model of overall self-being (see Yates 1997, 51). The *Hengxian* supplements this with the more abstract concept of "natural activity," indicating the natural unfolding of life in the world. "If all the activities of beings in the world remain free from constraints and dependencies, they can be natural activities"—referring to all beings living, working, and functioning in perfect self-being. The text also uses the concept of vitan energy to speak about the self-generation and self-activation of beings—closely relating to ideas of self-transformation, inherent stability, and the like.

The Daoist idea of self-generation further leads to an entire string of natural or "self" activities: self-transforming, organizing, ordering, stabilizing, clarifying, simplifying, submitting, generating, balancing, flourishing, righting, giving, examining, completing, acting, producing, and so on. All these are terms combined with "self" in various ancient documents, showing just to what degree Daoists emphasize the spontaneous, natural, "self" order of things. To them, the myriad beings going along with the nonaction, softness, and weakness of Dao means that they naturally unfold and of themselves act in the world—expressing how they function in perfect harmony with their unique needs and desires.

In the universe, in addition to giving all things and beings the ability to satisfy themselves naturally, Dao also has a yet higher requirement that reaches beyond them: in close mutual interaction, all things and beings should separately satisfy themselves and completely fulfill their fundamental nature, striving for as long as there is anything that remains unfulfilled. The *Laozi* says of Dao,

> It operates everywhere and is free from danger. . .
> Great, it functions everywhere.
> Functioning everywhere, it is far-reaching.
> Far-reaching, it returns [to the origin]. (ch. 25)

The *Huangdi sijing* similarly notes, "It operates everywhere and is always beginning. One can think of it as the mother of the myriad beings. Emptying and filling, it is their course [of life]." In other words, it is the continuous, ubiquitous, ongoing order of the universe (Wang 2007, 36-38).

At the same time, Daoists also acknowledge that the myriad beings in actual reality are not in complete harmony and cooperation with Dao. In their post-creation existence, they tend to manifest abnormality and alienation. To avoid and/or correct these, Daoists acknowledge that they must return to Dao. As Guodian A has,

> Once there are regulations, there are names.
> Once there are names, know that it is time to stop.
> Knowing when to stop, one can be free from danger. (ch. 32)

> They transform and develop a desire for activity.
> At this point, be careful to maintain the simplicity of the nameless.
> Also, know when it is enough.
> Knowing when it is enough, they can be still.
> From here, the myriad beings realize their inherent stability. (ch. 37)

Whether described in terms of "knowing when to stop" or "maintaining the simplicity of the nameless," this expresses the demand that all things and beings function in the way that is right for them, satisfying their unique nature and reaching a state of stability and inner peace—effected in concrete reality through the agency of the sage king. Based on the Daoist understanding that "Dao resting in nonaction and the myriad beings finding their self-being" is the ideal state of life in the world, the notion of return means that things and beings, faced with the constantly changing reality of life, still maintain a close connection to Dao and nature. This feature is characteristic of Chinese philosophy as a whole: it sees organic integration from the universe on down to each individual human being, expressed in notions of pervasive continuity, correlative thinking, and ideas of the harmony of nature and humanity (Graham 1986; 1998, 1-57; Du 2004, 86-91).

Human Society

Within this overall, broad philosophical system, then, Chinese thinkers also recognize the need to account for complexity and diversity, but they do so differently. Most Daoists strongly emphasize the "unity of heaven and humanity," Zhuangzi more so than Laozi and Huang-Lao thinkers, from whom he generally tends to diverge. For example, Zhuangzi is against government and advocates the dissolution of administrative structures. He focuses largely on the individual and proposes that each person should find his or her unique harmony with self-being and heavenly virtue, maintaining a consistent state of high-level continuity.

Huang-Lao Daoism, on the other hand, focuses strongly on community life and centers largely on rules and regulations, forming a school of thought concerned mainly with society and government. Gu Liya and Shi Huaci describe it as purposive or instrumental Daoism to flush out its tendency toward social realism and strong government control (Shi 2004, 244-45). Huang-Lao appears dominantly in both the *Shiji* 史记 and the *Hanshu* 汉书 ch. 3, which criticize its representatives for "releasing all action, to the point where they want to do away with rites and learning, abandon benevolence and righteousness, and rely solely on clarity and emptiness to govern the world"—ideals that closely match Zhuangzi. The utilitarian and practical political tendencies of Huang-Lao easily create the false impression that it is only concerned with things in the real world, its urge toward naturalism and ideas of the unity of heaven and humanity often making modern readers uncomfortable.

Scholars today get their ideas about this school largely from the literary section of the *Hanshu* ("Yiwenzhi" 艺文志) as well as the so-called arts of Dao used by Daoists in general and Huang-Lao masters in particular (Zhang 1982, 302-03). In fact, from the perspective of cosmology and cosmogony, Huang-Lao and Zhuangzi have a lot in common. They diverge significantly with respect to the role of the individual and society, having different ideas of what exactly "unity of heaven and humanity" means.

That is to say, while Zhuangzi wants to reduce the impact of society and the individual as much as possible to let heaven shine forth more strongly, Huang-Lao thinkers propose the realization of natural order in society through rules and structure. They hope to establish a large, vibrant, and effective social and political system in close connection with the natural order of the universe, using, as they say, "the Dao of heaven to illuminate human affairs." In this manner, they express a cosmology of natural integration and inherent continuity. Excavated manuscripts, and especially the *Huangdi sijing*, illuminate their perspective, helping us understand their cosmology and social vision in greater detail.

Huang-Lao Daoism connects the natural universe with human society in mutual unfolding. It sees continuity with nature in two dimensions: through Dao and oneness as the root of government, and through the myriad (human) beings as the core of society. These two forms of continuity connect to nature on two different levels and work with different models of its function. However, in their overall structure they reflect the expansion of the natural order in human society and express their inherent unity, high-level cooperation, and mutual interaction.

On one level, this continues Laozi's ideal of rule by nonaction, through which alone the most straightforward and most efficient government can be realized. However, as noted earlier, nonaction is originally a model of how Dao works, applied in Laozi to all aspects of human society. "If lords and kings can hold on to this, the myriad beings will self-transform," he says (ch. 37), expressing how the Dao of nature turns into the laws of the sage king. Huang-Lao thinkers inherit this thinking and similarly see nonaction as the root of Dao and rulership as its practical application. As the *Shida jing* says, "Forms constantly determine themselves: for this reason I am more at rest. Affairs constantly take care of themselves: for this reason I am in nonaction" (ch. 15; see Yates 1997, 152). The "I" in this statement, like the first person in the *Laozi*, refers to the ruler. Nonaction, on the other hand, as principle is not a concrete form of government. To make this a more widely applicable mode, Huang-Lao thinkers adopt the Legalist concept of law into their Daoist philosophy of rulership and add it to the foundation of Dao. They take Laozi's vision of the continuity of nature and humanity one step further and expand it to include the continuity between human laws and natural law.

In principle, Laozi does not denigrate law and institutions; he is just against harsh decrees and excessive regulation. Still, law and institutions are originally quite beyond his purview; he proposes fundamental principles of government and does not deal with specific operational systems. Huang-Lao thinkers integrate an institutional dimension and legal norms into Daoism, connecting it with concepts of nature, Dao, and oneness as their ultimate foundation. Here, then, Dao becomes "natural law" (Hu 1998, 198-225; Gao 1994, 212-54), with human laws as its extension. Combining the Dao of traditional Daoism and the law of Legalism, Huang-Lao thinkers thus create the new concept of the "law of Dao" or "Dao as law" (*daofa* 道法), the title of chapter 1 of the *Chengfa* in the *Huangdi sijing*.

The text here says, "Dao gives birth to law," clearly designating law as the direct product of Dao. This means, law in human society is the direct effect of the order and principle of the universe. As Shi Huaci says, "When the *Huangdi sijing* says that 'Dao gives birth to law,' it speaks of law as the force that rules both the natural world and also the entire complex of human culture. It is a natural process that evolves from nature into society" (2004, 257).

However, Dao cannot give birth to human laws directly, but must work through high-level humans. As the "Daofa" has it, "Those who hold on to Dao produce law and do not venture to transgress it; establishes law and do not venture to oppose it" (Yates 1997, 51). In other words, law is established by those who have a strong "hold on Dao"; ideally, it relies on Dao, which makes it "law of Dao." The *Guanzi* similarly says that all "constitutions, laws, and institutions must be patterned on Dao" ("Fafa" 法法), clarifying this point. The law of Dao can preserve the inherent order of the cosmos and the myriad beings because it reflects a high level of unity and pervasive applicability—as the *Hanfeizi* says, "The law of Dao is always whole and complete."

The Role of the Ruler

Huang-Lao thinkers use the concept of oneness to express this idea. According to traditional Daoist thought, Dao uses law to control the masses of creatures in the universe: the one managing the many. In true accordance, the ruler in human society uses oneness as expressed in the law of Dao to control the masses of the people. When Laozi speaks of "holding on to oneness" or "to Dao," he really means "maintaining nonaction" and has no intention to see this manifest in particular laws or institutions. Once Huang-Lao Daoism arises, however, the ruler holding on to oneness, Dao, or nonaction comes to mean holding on to the oneness of the law. By mastering and applying oneness in the shape of universal and uniform laws, the ruler can attain nonaction. Seen from a different perspective, this means he uses laws as his central mode of governing precisely because he pursues the principle of nonaction.

The *Daoyuan* adds to this,

> Be highly trustworthy and free from affairs,
> And the myriad beings will rally around you.
> Divide them according to their lots,
> And the multitude of the people will be free from contention.

Give to each according to their names,
And the myriad beings will be self-stabilizing.

Do not labor on behalf of order;
Do not be idle because of disorder.
Be vast and expansive,
And troublesome tasks will not reach you.
Be deep and subtle,
And searching will not find you.

Now be oneness and do not transform,
Attain the root of Dao.
Grasp the few to know the many,
Attain the essentials of affairs.
Lay hold of the correct to correct the irregular.

First know highest antiquity,
Later [can] be quintessentially bright.
Embrace Dao and hold on to measures,
And the world can be unified.
Observe highest antiquity,
And discern everywhere where it can be used.
Search into matters before they are nothing;
Control them when you have the means. (see Yates 1997, 177) [6]

Here the phrases "be oneness" and "root of Dao" as well as "embrace Dao and hold on to measures" and "the world can be unified" are particularly illuminating. Daoists generally exclude all foreceful, "make do" modes of governing and strongly emphasize the practice of nonaction and application of Dao in simplicity, softness, and weakness as the best way of the ruler to establish law.

The *Hengxian*, too, asserts that "forceful" means "doing much" and says that "unless one conceals oneself, one cannot possibly be as one truly is," encouraging a modest and self-effacing behavior in favor of forcefulness. Here

---

[6] Reading this passage, most commentators take the entire first part, down to "Be oneness," as one paragraph, but this is not correct. Examining the structure more closely, it becomes clear that the sentences following it are parallel and match each other, which means the meaning should also match. "Be oneness and do not transform" is in fact part of the second paragraph, which should be clearly set apart. While the first part speaks about the best way the ruler should practice Dao and effect the laws in the world, the second part illuminates the best actions to take for the common people.

reasonable and appropriate doing or action all involve honoring the law while going along with Dao. "Honor all the world does without interference or negation," which means avoiding any opposition to the constant Dao as it works through all aspects of life. Laozi expresses the same idea when he says, "He who does not lose his place [with Dao] can endure" (ch. 33). There is, then, a strong continuity between the nonaction of Dao and that of the ruler, between the law of Dao and the laws of humanity.

Another aspect Huang-Lao thinkers emphasize is the continuity between the self-being of the myriad beings and that of humanity. As Laozi says, humans are one among the myriad beings, which means that their life and activities match to a large degree. Now, if the self-being of the myriad beings is in fact the nonaction of Dao, this means that human self-being is also the nonaction of Dao as modeled by the sage kings. Vice versa, this also means that Dao in the greater universe follows the self-being of the myriad beings, and the ruler in human society follows that of the people.

Huang-Lao thinkers use specifically Laozi's notion of "following," as manifest in expressions such as "stillness follows Dao" and "following along." From here, they require that the ruler honor and follow the self-being of the people. As Guodian C has, "They [the rulers] accomplish their task; they complete their work. Nevertheless the people say, we are just natural," which in the transmitted version reads, "the people say that they simply follow nature" (ch. 17). Either one refers to the activities of humans as one among the myriad beings as they match the nonaction of the ruler. Guodian A has,

> Therefore, the sage says:
> I do not engage in affairs,
> And the people are self-prospering.
> I let go of all action,
> Aand the people are self-transforming.
> I relax in stillness,
> And the people are self-righting.
> I prefer being free from desires,
> And the people are self-simplifying. (ch. 57)

All these are ways to allow the myriad beings to live in harmony according to their particular inherent nature; humans as one of these beings, too, should follow their natural abilities as given by Dao. Laozi's understanding of the inherent nature of all beings applies also to humans. It is the driving force of all human life and activity, their pervasive self-being rooted deep within. However, he never clarifies what exactly it means for the people to be self-transforming, self-righting, or self-simplifying.

Self-Interest

Huang-Lao thinkers fill this gap, understanding inherent nature in people to be their spontaneous feelings, their inner urge to "pursue benefit and avoid harm," "love life and hate death." In this context, they introduce the concept of "self-acting," i.e., acting to one's own advantage, as part of their understanding of human inherent nature. They recognize that thinking of oneself first is a fundamental human trait and believe that humans are essentially self-serving creatures. Accepting this as natural and not necessarily bad, they never reach the point where they want to change people's nature. On the contrary, they confirm that self-interest is the underlying driving force of all human behavior and insist that rulers work with it to establish a well-functioning social and political order. Rulers should closely follow the inherent self-interest of the people and strive to satisfy it as much as possible.

This, however, is an oversimplification of human feelings, emotions, and natural urges. In fact, human beings do not only think of material self-interest and personal well-being but also pursue the life of the spirit and sacrifice much for ethical and moral values. We can phrase the issues in terms of the Scottish philosopher Adam Smith (1723-1790). In his *The Wealth of Nations* (1776), he focuses on human self-interest as the driving force of all economies, noting that "every individual intends only his own gain." On the other hand, in his *A Theory of Moral Sentiments* (1759), he insists that moral judgments often supersede self-interest. He thus contrasts economic man and moral man as two different dimensions of human life.

For Huang-Lao thinkers, this means that their contention of the effectiveness of legal rewards and punishments rests firmly on their belief that the actions of all beings, including humans, are naturally determined by self-interest. The oneness of the law can control the multiplicity of the people because it addresses the underlying oneness of their joint self-interested nature. Zhuangzi notes that certain thinkers such as Shen Dao 慎到, Tian Pian 田骈, and Peng Meng 彭蒙 emphasize "making all beings equal" (ch. 33), i.e., unifying them in oneness by acknowledging that they all share a single common nature that can be corrected and streamlined.

For humans specifically, the Huang-Lao vision of equalizing and unifying does not mean to eliminate all distinctive individual traits. Rather, it means to find an underlying unity, a fundamental sameness, a oneness that pervades all variety precisely in the vast multiplicity of individuals, "in the thousands of distinctions and millions of differences." This, then, they can measure and regulate. As the *Lüshi chunqiu* says, "If you equalize all differences among the myriad beings, then all of them, whether ignorant or knowledgeable, skillful or clumsy, will exhaust their strength and work to the best of their ability, as if they all came from the same cave" ("Buer").

Huang-Lao thinkers have no desire to change people's inherent nature—they do not expect the ignorant to turn into fonts of wisdom. Their ideal is that human beings live according to their essentially equal natural temperaments and unique abilities in life. Therefore, they take Laozi's idea that the ruler should follow along with the people's self-being and develop it into the use of laws that would allow individuals to realize their natural inclinations. That is to say, Daoists in general have the tendency to strive for a return to inherent wisdom, but the question is, what wisdom exactly do specific Daoists wish to recover?

They generally agree that rulers should not use their personal wisdom and charisma to govern by intentional action. Instead, they should go along with the people's inherent temperament and unique aspirations and execute a government by law. As Shi Huaci explains it, "Based on this, we can clearly understand why Shen Dao recommended 'ridiculing the wise' without, however, negating their value. For him, the social order cannot rely on error-prone individuals with their particular ethics and personal judgments. Instead, it must rely on the law as determining cultural order, the law itself being the representative of Dao on earth" (2004, 254).

Both Laozi and Huang-Lao Daoists believe that the law of Dao is pervasive in the universe and essential to the establishment of order. They trust the natural inclinations of human beings. The most efficient government has to grow from unified law; it cannot rely on the limited wisdom and personal charisma of any one individual. For exactly the same reason Aristotle insisted that rule by law is superior to that of wisdom.

Conclusion

The newly excavated manuscripts offer important insights into early Daoist thought, especially when combined with a new reading of the transmitted texts. At this time, it is not only possible but imperative to reevaluate early Daoist views of cosmogony and human development. The most important sources in this context are various securely dated manuscripts, notably the *Laozi*, *Huangdi sijing*, *Hengxian*, *Taiyi shengshui*, and *Fanwu liuxing*. They provide new testimony to early Chinese conceptions of nature and the universe, created as they were in an essentially Daoist mode.

Ancient Daoists proposed various theories of cosmic unfolding as well as different descriptions of the primordial state, opening the path to a plethora of competing models. They offered various interpretations of the original nature of the myriad beings and looked differently at the relationship between Dao and beings, simplicity and complexity, the one and the many, Dao and virtue, etc. They used concepts such as the nonaction of Dao and the self-being of the myriad beings to describe the universe and all living beings and illuminate their inherently good order. Notions such as weakness as a function of Dao and the myriad beings having certain inherent natural characteristics open a new and better access to their thinking. Ideas of the nonaction of the sage and the self-being of the people providing the foundation of a human order that fundamentally reflects that of the greater universe involve concepts of oneness, law, and human inclinations essential to a highly mature political philosophy. These not only illuminate the arising and flourishing of ancient Chinese natural philosophy and cosmology, but also elucidate its internal reasoning and visions of humanity as the direct expansion of universal forces.

Chapter One

Hengxian

Stages of Cosmic Unfolding

From the *Huangdi sijing*, unearthed at Mawangdui, to the *Taiyi shengshui* and the *Hengxian*, excavated at Guodian, the texts reveal a yet unheard-of richness and diversity of late-Zhou Daoist cosmology. From the early disciples of the old master to the nine *Laozi* versions extant today, from Laozi to Zhuangzi, the exact lineage and development of Daoist thought remain shrouded in mystery (Li 2006; Wang 2006, 494-95). The differences between the *Hengxian* and the *Huangdi sijing* may well reflect major philosophical developments of the late Warring States period, filling in the gap between Laozi and Zhuangzi and thereby offering a better understanding of a highly complex philosophical and social situation.

The manuscripts have their limitations, but as sources, they are invaluable. Both Guodian texts focus strongly on the emergence and unfolding of the universe, explicating specific points of cosmogony based on essentially the same philosophical background, Laozi's cosmology. That is to say, they outline new visions of cosmic emergence, allowing a glimpse into the theories and arguments of late-Zhou Daoist cosmology.

Some key points stand out. For example, in the *Laozi*, the key cosmological concept is Dao, but neither the *Hengxian* nor the *Huangdi sijing* make much of it, using instead Constancy Before and Great Oneness to describe the original state or prime mover of the universe. The only place where the *Hengxian* speaks of Dao is in the line, "Once the Dao of heaven is laid out," and even here "Dao" may well be a variant for "earth," so the line reads, "Once heaven and earth are laid out" (Li 2003, 296; Pang 2004, 22). The *Shida jing* in the *Huangdi sijing* similarly has, "Once the Dao of heaven is established, earth and all beings come to be." "Laid out" and "established" are close in meaning, which suggests that the *Hengxian* actually means "Dao of heaven" and not "heaven and earth."

Still, this is the only place where the text uses the term "Dao." Why, then, we must ask, does the *Hengxian* avoid "Dao"? Why does the *Taiyi shengshui* not present a clear analysis of its role in cosmic origination? Most Daoist texts of the period, including the *Zhuangzi*, *Guanzi*, *Wenzi*,

Huainanzi, and even parts of the *Huangdi sijing* acknowledge Dao as the root of creation and use it as a fundamental philosophical concept. Why is the *Hengxian* so different?

Many scholars have examined and discussed the text,[1] but these issues remain. Its unique philosophical vision needs to be examined in more detail; its development of Daoist metaphysics and understanding of the relationship of heaven and humanity must be explored more fully.

Constancy Before

The *Hengxian* begins,[2]

> In Constancy Before, there is no material existence. There is simplicity, stillness, and emptiness. Simplicity is Great Simplicity; stillness is Great Stillness; emptiness is Great Emptiness. It fulfills itself without repressing itself.

The fundamental cosmological concept here is Constancy Before, which many scholars equate with Dao (e.g., Li 2003, 288). However, I think the text intentionally avoids Dao and uses a new term instead to name the root source of the universe. The *Daoyuan* in the *Huangdi sijing* has the expression "constant nonbeing" (*hengwu* 恒无), but it still uses Dao as its essential concept. The idea here is not the same as that expressed as Constancy Before. If the latter were in fact the same as Dao, it would overshadow the uniqueness of the *Hengxian's* cosmology.

Qiu Xigui argues that the word "constant" (*heng* 恒) should be read as "ultimate" (*ji* 極). He says that the plain graph *geng* 亘 is really *han* 亟, the key part of the word "ultimate," reflecting a substitution that occurs commonly in the bamboo manuscripts. The *Shuowen* glosses "ultimate" as "rafter," i.e., the central beam in a house. "It means stretched long; the highest and longest beam is called 'ultimate'." In addition, the term carries the impli-

---

[1] Earlier studies of *Hengxian* cosmology include Asano 2005; Chen 2004; Ding 2005; Li 2004; Li Rui 2004; Liao 2004; Lin 2005; Pang 2004; Takeda 2005; Zeng 2006a.

[2] This translates the first half of the text, following the editions in Li 2003 and Pang 2004. The text is fragmentary and consists of thirteen bamboo slips, here indicated by numbers in parentheses. The English version follows Brindley et al. 2013.

Li Ling divides it into four parts: slips 1-4, 5-7, 8-9, and 10-13. However, the discourse is continuous and any division is artificial. Pang Pu sees a slightly different division, which most scholars have come to accept. He places slips 8-9 after slip 4, then has 5-7, followed by 10-13.

cation of "central" and "standard." Qiu thinks that the most original and initial state of the universe should be called Ultimate Before makes a lot of sense, as does the combination "constant energy."

He further argues that, when the *Daoyuan* speaks of "the initiation of constant nonbeing," the term "nonbeing" should be read "before," so that the phrase really speaks about Constancy Before and matches the *Hengxian*. The manuscript *Yizhuan* 易传, moreover, has the phrase, "The changes have great constancy" rather than "great ultimate," but to Qiu "constancy" here is a copying error and also should really be read "ultimate." Altogether, therefore, he sees both etymology and meaning as suggesting that "constancy," written *geng*, is in fact "ultimate," a highly appropriate term for the state of universal beginning (Qiu 2008, 1-16).

I beg to differ on various points. First, while *geng* and *heng* are commonly interchanged in pre-Qin writings, both *Laozi* manuscripts use *heng* and the transmitted edition reads *chang* 常, changed in the Han due to the avoidance of the taboo of Emperor Wen's personal name, Liu Heng (r. 180-157). *Heng* is also central in the *Huangdi sijing*. Only few places suggest that *geng* is a substitute for "ultimate," and the *Yizhuan* passage in question is unique in that it contains a copying error (see also Rao 1993, 6-19; Nakajima 2005, 89-92).

Second, there is no reason why the *Daoyuan* "nonbeing" should be read "before," the phrase makes perfect sense as written and echoes the general Daoist emphasis on nonbeing, as is also manifest in terms like nameless and formless, used in the *Daoyuan*, "Before there is form, all greatly revolves in the nameless."

Third, the full phrase in the *Daoyuan* has, "At the initiation of constant nonbeing, all was merged in great emptiness," and later says, "Empty and same, all was in oneness, constant oneness, nothing more." Here "constant oneness" echoes expressions like "constant Dao" or "constant virtue" and there is no reason to read any of them as "ultimate." Similarly, the expression "constant energy" in the *Hengxian*, although written with the graph *geng*, is perfectly reasonable and there is no need to read it as "ultimate energy," which, moreover, in the overall cosmological model, is not at the beginning of creation, so that "ultimate" really makes no sense. In most places, *geng* substitutes for *heng* and is later replaced by *chang*, retaining the meaning of "constancy" and not "ultimate."

Fourth, the original meaning of "constant" is "eternal" and "ongoing." The *Yijing* has it as a hexagram, "Duration." It says,

> Duration means being eternal . . . in Dao. The course of heaven and earth is constant and eternal in its operation and never stops. Movement in any direc-

> tion will be advantageous: it ends and begins again. The sun and the moon attain heaven, and thus are able to radiate eternally. The four seasons change and transform, and thus are able to complete things eternally. The sages rest eternally in Dao, and the world is transformed and complete. Observing their constancy, the natural tendencies of the world and all beings become obvious. (Hexagram 32; translation based on Sung 1971, 140; see Wilhelm 1950, 126-27)

In other words, "constant" here means "eternal" and signifies the ongoing course of the heavenly bodies and natural cycles. Both "constant" and "before" refer to temporal phenomena, but "before" here is not an ordinary ranking in a sequence; rather, it indicates the eternal or primordial state before creation, the starting point and initial glimmer of the entire universe.

Pang Pu interprets it as Ultimate Before and says that Constancy Before is "the absolute first, the most initial of all initiation, quite similar to what Qu Yuan 屈原 in the *Tianwen* 天问 section of the *Chuci* 楚辞 calls the "initiation of Great Antiquity" (2004, 21), a temporal starting point of creation. In this sense, Constancy Before matches Great Oneness in the *Taiyi shengshui*, the "initiation of constant nonbeing" in the *Daoyuan*, and even Great Initiation of the *Zhuangzi* (ch. 12). It is, moreover, the point or origin where, as the *Hengxian* says, "there is no material existence," when all is still nameless and formless.

Cosmic Beginnings

In terms of overall structure, the fundamental element that gives rise to the myriad beings in its primordial purity is their ultimate source and origin. From the perspective of the unfolding of life, the myriad beings evolve from the root source of the entire universe. In Chinese philosophy, this root source is often linked to Dao or vital energy as, for example, in expressions like "essential energy" or "primordial energy." Once the universe has emerged, i.e., taken birth and reached a certain level of completion, it becomes the factor that "gives birth to the born" and "gives birth to completion." In addition, philosophers also discuss it in terms of "that which gives birth to the born" and to "completion." For example, the *Liezi* 列子 says,

> There are the born and the unborn, the changing and the unchanging. The unborn can give birth to the born, the unchanging can change the changing. The born cannot escape birth; the changing cannot escape change. Therefore, there is constant birth and constant change. (ch. 1; Graham 1960, 17)

That which can "give birth to the born" and "give birth to completion," then, in the *Hengxian is* "self-birthing and self-working," is the root of all existence, i.e., the "state of constancy." In terms of universal emergence and development, what gives birth is temporally earlier, what came before everything else, the original initiation of the universe.

Human beings, full of curiosity about themselves, pursue questions about their origin; full of curiosity about nature, they ask where the myriad beings came from. This strong awareness of the origin, the quest for ultimate belonging is a unique characteristic of humanity.[3] In the world's religions, these find expression as ancestor worship and belief in various deities; in philosophy, they emerge as different ontologies and origin theories. As Leo Strauss notes, the most fundamental quest of philosophy is not only concerned with questions of existence, but also with finding a consistent and irreducible understanding of its origins. Many different theories and hypotheses all rest on the basic premise that all things have a cause. If some say that in the beginning all arose from chaos, a state when there was nothing, no material existence, this is unreflected talk (1953; 2003, 90).

This is highly reminiscent of Qu Yuan's *Tianwen*, where he asks, "Who passed down the story of the far-off, ancient beginning of things? . . . How did yin and yang come together? How could they originate and transform all things that are by their commingling" (Hawkes 1959, 127). Similarly, in the *Liezi* chapter "Tangwen" 汤问, King Tang of Shang asks Xia Ge, "Have there always been things?" He replied, "If once there were no things, how come there are things now? Would you approve if the men who live after us say there are no things now?" (ch. 5; Graham 1960, 94). Zhuangzi. too, says,

> There is a beginning. There is a not yet beginning to be a beginning. There is a not yet beginning to be a not yet beginning to be a beginning. There is material existence. There is nonbeing. There is a not yet beginning to nonbeing. There is a not yet beginning to not yet beginning to nonbeing. (ch. 2; Watson 1968, 43)

In the *Hengxian*, the root and origin of the universe is Constancy Before. It matches the idea of Dao in the *Laozi*, Great Oneness in the *Huangdi sijing*, Great Initiation of the *Zhuangzi*, and Great Change in the *Liezi*—which places Great Initiation and Great Beginning after this—as well as various other early notions.

---

[3] In this context, human beings make the distinction between themselves and the myriad beings, between self and society, leading to a difference in awareness of themselves and of others as well as to distinctions between humanity and the animal kingdom, between individual and the masses.

Nonbeing, Simplicity, and Stillness

Now, if Constancy Before indicates the primordial state and original beginning of the universe, the text should make clear just what kind of phenomenon it is. Indeed, it speaks of it as having "no material existence" but being "simplicity, stillness, and emptiness." "No material existence" means that there is nothing there, a state of "nonbeing" (*wu* 无). According to Daoist metaphysics, the distinction between material existence and nothingness is that the former has form and manifests in all kinds and sorts of concrete things, often also called "vessels" (*qi* 器). Nothingness is understood generally as being formless, a state of non-differentiation or chaos that is imageless and nameless (Wang 2001, 131-68).

The *Hengxian* acknowledges the original state of the universe as nonbeing, however, this indicates a state before forms, images, and names; it is not a state of absolute nothing with no existence of anything at all. If there were nothing at all, that is, this could not possibly give birth to something that manifested in forms, images, and names. When the 3rd-century commentator Wang Bi 王弼 or other Daoist thinkers use nonbeing to describe the root source of the universe, then, they do not mean "nothing at all."

Similarly, the *Laozi* contrasts its concept of Dao as universal origin with beings that are "formed vessels" and characterizes it as not being formless and nameless, but not absolutely nothing. Laozi's Dao is a "thing" (*wu* 物), and this thing manifests as "images" (*xiang* 象). However, while these are not ordinary images but "great images," they are yet very much real. Zhuangzi says, "In Great Initiation, there was nonbeing, no material existence, no names" (ch. 12), closely matching the vision of in the *Hengxian* even if their expressions are different. Zhuangzi also says, reflecting on the death of his wife, "I looked back at her beginning and the time before she was born—not only the time before she was born, but the time before she had form" (ch. 18; Watson 1968, 192). This again characterizes the original state of the world as formless.

The *Hengxian* further describes the original state as characterized by "simplicity, stillness, and emptiness." These, too, are not ordinary qualities but all called "great." The word "simplicity" here (*pu* 朴) is somewhat controversial. Li Ling interprets it to mean "substance" (*zhi* 质) while allowing for a sense of "simplicity" in its meaning (2003, 288). Li Xueqin reads it as "wholeness" (*quan* 全), which changes the expression Great Simplicity to Great Wholeness (2003, 81). The latter occurs in the *Zhuangzi*: "I do not know the great wholeness of heaven and earth" (ch. 21). However, this does not refer to the origin of the universe but is a reflection on the vastness of heaven and earth and their quality of encompassing the myriad images.

Zhuangzi further speaks of "the greatness of heaven and earth, the multiplicity of the myriad beings" (ch. 19), while the "Zhongyong" 中庸 chapter of the *Liji* 礼记 notes, "The greatness of heaven and earth: human beings are deeply moved by it." In antiquity, the expression "heaven and earth" commonly referred to the largest and most extensive entity, their "great wholeness," expressing their quality of encompassing the myriad beings. Based on all this, I think the reading "simplicity" and "Great Simplicity" is quite appropriate.

Another term with an alternate reading is "stillness" (*jing* 静). Li Xueqin reads it "clarity" (*qing* 清); Liao Mingchun matches this by reading "Great Stillness" as "Great Clarity" (Li 2003, 81; Pang 2204, 83-84). However, to me the original reading is still the most meaningful. All three terms, "simplicity," "stillness," and "emptiness," occur in the *Laozi*—which uses "stillness" more often than "clarity"—and have played an important role in later Daoist thought. In the *Laozi*, they describe the inherent state and major characteristics of Dao, give expression to its realm and perfection, yet they also indicate qualities of concrete, everyday things as they emulate Dao. Since these three, as applied to Dao, are already at their highest possible level, the *Laozi* has no need to add the epithet "great," but the *Hengxian* does, emphasizing their importance at the very cusp of universal origination. The epithet "great" (*tai* 太), moreover, can be understood to indicate the utmost, highest, or most perfect level of being, so that the three qualities described as "great" are really "perfect" (*zhi* 至).

Self-Doing

The *Hengxian* continues,

> There is not yet heaven and earth; there is not yet working, progression, emergence, or birth. Muddled and murky! All is still and homogeneous. There is not yet light, not yet teeming life.

This means that in its state of origin, the universe is uniform and stable, unchanging, unmoving, unreflecting, in a "state of even constancy." Reflecting the most common way Daoists look at cosmic beginning, this is also borne out in the *Wenzi:*

There is a being, chaos yet complete,
Born before heaven and earth.
Pure image, without form,
Deep and obscure!
Muddled and murky!
One cannot hear its sound. (ch. 1; "Daoyuan" 道原)

While this has some resonance with the *Hengxian* description, the latter is closer to the *Daoyuan*:

All was merged in great emptiness.
Empty and same, all was in oneness,
Constant oneness, nothing more.
Misty and blurred,
There were neither light nor darkness.
Spiritual and subtle, yet it filled everywhere;
Quintessentially still, it was not luminous. (see Yates 1997, 173)[4]

This describes the self-birthing and self-originating universe as outlined in the *Hengxian*, a state that is naturally self-sufficient, which the text describes as "self-fulfilling." However, the universe cannot just remain in this state; as a state of origin and beginning, it has to give birth to things and raise them. It thus "does not repress itself." Li Ling explains this as not subduing itself, not pushing itself down. The *Laozi*, too, is much in favor of "self" workings, as is evident in its many compounds: self-being, self-transforming, self-balancing, self-correcting, self-aware, self-purifying, self-enriching, and more. These expressions all describe the inherent qualities of the universe as it, after giving birth to the myriad beings, lets them unfold without controlling or pushing them.

It gives birth to them,
But does not own them;
It acts on them,
But does not make them dependent;
It grows them,
But it does not master them. (ch. 51; Chan 1963, 163)

---

[4] Zhang Heng 张衡, in his *Lingxian* 灵宪, says something quite similar, "Before Great Plainness, all was murky clarity, mysterious stillness, serene quiet, and dark silence—impossible to imagine. Inside it, only emptiness; beyond it, only nothingness. This state went on for a very long time. We call it the infinite. It is the root of Dao." See Liu 1987, 3215.

The text also outlines various conditions with regard to "self" that one should avoid: consider oneself right, see oneself as great, fight for oneself, have self-pity, give birth to oneself, value oneself highly, and so on. Laozi's Dao strongly values softness, weakness, stillness, simplicity, emptiness, non-action, and non-contention. The *Hengxian*, too, uses a number of terms relating to "self," including self-birthing, self-working, self-fulfilling, self-happening, and self-reproducing. In negative expressions, besides "not repressing itself," it also has "not being like itself," thus supplementing its overall positive lifestyle description. Terms like forbearance, repression, and restraint are typical for Confucian thinkers; Daoists, in contrast, prefer going along with life in self-being, free from control and constrictions. Constancy Before as the infinite state of cosmic beginning, is like that: essentially self-sufficient, flowing along with its own being, and free from repression.

Dao as Source

The *Hengxian* next says,

> Space works. Once there is space, there is energy; once there is energy, there is material existence; once there is material existence, there is a beginning; once there is a beginning, there is the passage of time.
>
> There is not yet heaven and earth; there is not yet working, progression, emergence, or birth. Empty, still, and as though one. Muddled and murky! All is still and homogeneous. There is not yet light, not yet teeming life [see Zeng 2006b, 114].
>
> Energy is self-birthing. Constancy categorically does not give birth to energy. Energy is self-birthing and self-working.
>
> The birthing of constant energy does not happen in isolation; there is something that takes part [in the process]. Space and constancy are both present. The same can be said of the process, during which space is born.

Generally, Daoist metaphysics consists of a close combination of ontology and cosmogony, with different thinkers placing their dominant emphasis on different aspects. Laozi tends to focus on Dao as the pervasive source of the universe, but he also recognizes it as the root and cause of the myriad beings, their ultimate essence. Similar metaphysical notions, obviously under Laozi's influence, also appear in the *Huangdi sijing* (*Daoyuan*) *Guanzi* ("Xinshu shang"), *Zhuangzi* (ch. 6), *Wenzi* (ch. 1), and *Huainanzi* (ch. 1). They all speak of Dao as the root cause of the myriad beings and acknowledge that they rely on, and are pervaded by, Dao in their original characteristics and essential rightness. Still, there are subtle differences.

For example, the *Daoyuan* in the *Huangdi sijing* strongly sees Dao less as the source of the universe and more as the root and essence of the myriad beings. It says,

> Therefore, it never depends on existence;
> Never relies on any of the myriad beings.
> Therefore, it does not possess form;
> Is immensely pervasive yet nameless.
>
> Heaven cannot cover it;
> Earth cannot hold it up.
> So small, it can bring smallness to completion;
> So large, it can bring largeness to completion.
>
> It fills up all within the four seas;
> And embraces all beyond them.
> In yin, it never rots;
> In yang, it never scorches.
> It takes oneness as its measure and never changes,
> And is right there even for crawling insects.
>
> Birds attain it and fly.
> Fish attain it and swim.
> Wild animals attain it and run.
> The myriad beings attain it and live.
> The hundred affairs attain it and are complete.
>
> All men use it,
> But no one knows its name.
> All men employ it,
> But no one sees its form.
>
> Oneness is its appellation;
> Emptiness is its dwelling.
> Nonaction is its element;
> Harmony is its function.
>
> For this reason, highest Dao is
> So lofty it cannot be scrutinized;
> So deep it cannot be fathomed.
> Clear and bright, no one can name it;
> Vast and large, no one can give it form.

It stands alone and is not paired with anything else,
None of the myriad beings can order it about.

Heaven and earth, yin and yang,
The four seasons, the sun and the moon,
The planets and constellations, primordial energy,
The wrigglers that walk and crawlers that move,
Even the plants that grow roots—

They all take life from Dao,
But never decrease it.
They all return to Dao,
But never increase it.

Hard and strong, it is never broken;
Soft and weak, it cannot be changed.
Subtle and fine, it can never be reached;
Far and distant, it cannot be passed.
(see Yates 1997, 173-75; Chang and Feng 1999, 197-98)

A similar point is also made in the *Zhuangzi*:

> Dao has its essential characteristic and its signs, but is without action or form. You can hand it down but you cannot receive it; you can attain it but you cannot see it. It is its own source, its own root. Before heaven and earth existed it was there, firm from ancient times. It gave spirituality to the spirits and gods; it gave birth to heaven and earth. It exists beyond the highest point, and yet you cannot call it lofty; it exists beneath the limit of the six directions, and yet you cannot call it deep. It was born before heaven and earth, and yet you cannot say it has been there for long; it is earlier than the earliest time, and yet you cannot call it old.
>
> Xiwei attained it and held up heaven and earth. Fuxi attained it and entered into the mother of energy. The Big Dipper attained it and from ancient times has never wavered. The sun and moon attained it and from ancient times have never rested. Kanpi attained it and entered Kunlun. Pingyi attained it and wandered in the great river. Jienwu attained it and lived in the great mountain. Huangdi, the Yellow Emperor, attained it and ascended to the cloudy heavens. Zhuanxu attained it and dwelt in the Dark Palace. Youqiang attained it and stood at the limit of the north.
>
> Xiwangmu, the Queen Mother of the West, attained it and took her seat on Shaoguang—nobody knows her beginning, nobody knows her end. Pengzu attained it and lived from the age of Shun to the age of the Five Dictators. Fu Yue attained it and became minister to Wuding, who extended his rule over the whole world; then he climbed up to the Eastern Governor, straddled the Win-

> nowing Basket and the Tail, and took his place among the ranks of stars. (ch. 6; Watson 1968, 81-82)

These passages explain Dao from different angles, focusing on its fundamental essence and limitless nature and illustrating how the myriad beings attain their fullness through Dao, living and continuing. Dao here is not primarily the source that gives them birth. The latter occupies central stage in other works, notably the *Hengxian*, *Taiyi shengshui*, *Liezi* (ch. 1), *Huainanzi* (chs. 3, 7), *Lingxian*, and more. Exploring the deep mysteries of the universe and of nature, these texts place a great emphasis on how the world and all beings come into being.

Properly speaking, the metaphysics of the *Hengxian* is a form of cosmogony and not ontology. Still, it is not the same as that of the *Laozi*, *Huangdi sijing*, and others. Laozi outlines the process from Dao to creation:

> Dao gave birth to one.
> One gave birth to two.
> Two gave birth to three.
> Three gave birth to the myriad beings. (ch. 42)

The *Taiyi shengshui* presents more stages, beginning with Great Oneness and ending with the cycle of the years (*sui* 岁); in addition to the expression, "give birth to" (*sheng* 生), it also uses the phrases "return to the pivot" (*fanfu* 反辅) and "recovering the mutual pivot" (*fu xiangfu* 复相辅). Some of this is reminiscent of the *Liezi*, which has,

> There was Great Change, there was Great Initiation, there was Great Beginning, there was Great Plainness. Great Change preceded the appearance of energy; Great Initiation was the beginning of energy. Great Beginning was the beginning of form; Great Plainness was the beginning of substance. Energy, form, and substance were together and not separate: thus, we call this state "chaos." Chaos means that the myriad beings were merged in confusion and not yet separated.
>
> Looking at it, one cannot see it; listening for it, one cannot hear it; pursuing it, one cannot attain it: thus we speak of "change." Change was without form or boundaries; it transformed and become one; one transformed and became seven; seven transformed and became nine; nine transformed and that ended it. From here it all returned and again became one.
>
> Oneness is the beginning of all form and transformation. The clear and light rose up to become heaven; the muddy and heavy sank down to become earth; the blended and harmonious energy became humanity. For this reason, heaven and earth contain their essence, and the myriad beings transform and take birth. (ch. 1; see Graham 1960, 18-19)

Space

According to the *Hengxian*, the universe evolves from space (*huo* 或) through energy (*qi* 气), material existence (*you* 有), and beginning (*shi* 始) to the passage of time (*wang* 往). It says, "Space works. Once there is space, there is energy; once there is energy, there is material existence; once there is material existence, there is a beginning; once there is a beginning, there is the passage of time." This outlines the gradual unfolding from nothingness to material existence, from nonbeing to being, using the expression "there is" (*you* 有) rather than "give birth to." The process, moreover, is one of ongoing transformation, leading to the emergence of one stage after the next.

The most difficult term to understand among these five levels is the first. What does the text mean by "space"? Li Ling notes, "From the context, it seems that the term indicates the status between the presence of some pure nothingness (Dao) and a sort of real existence (energy), but not either of them. Here 'space' is read as something existing; it includes the tendency toward diversification" (2003, 288). Constancy Before, that is to say, is not pure nothingness and is not the opposite of real existence. "Space" is a constant state of original formlessness that at the same time signifies the open presence of formless energy.

Cai Shuheng notes that the word used here, *huo*, originally indicates the domain of a feudal warrior and is later written with the "earth" radical, *yu* 域 (1983, 7). The *Shuowen* glosses it as a "combination of 'mouth' 口 and 'lance' 戈 to protect unity. This unity means land." This land, moreover, is a feudal possession. Later marked by a surrounding wall, the word forms the central part of the character for "state" (*guo* 國).

Pang Pu, too, explains "space" as "some kind of domain" (2004, 21). The word *yu*, moreover, appears in the *Laozi*: "There are four great things in the domain" (ch. 25). Both manuscript versions use the world "state" here, which supports the fundamental identity of the terms. Li Xueqin adduces the usage of "domain" in both the *Laozi* and the *Huainanzi* (ch. 3) to conclude that the term should really be "space" as opposed to "universe" (2003, 81-82). In fact, the texts suggest that "space" is the opposite of "shelter" (*yu* 宇). This is borne out in the *Mozi*, which notes that "long lasting" indicates a "protected time" and "shelter" means a "protected space" (Tan 1964, 227-29). Another early source that proposes a similar scenario is the *Huainanzi*. Here Dao begins in "empty vastness" (*xukuo* 虛霩); this gives birth to space-time (*yuzhou* 宇宙), from which in turn energy arises (3.1; Major et al. 2010, 114).

People today have a hard time imagining how "empty vastness" can give birth to time and space, and how the latter can give rise to energy. Traditional cosmologists generally acknowledge that space and time are material entities that have some kind of form but are free from concrete existence. This understanding echoes Kant's notion of time or space as a "thing in itself," an ineffable projected postulation that can only be described in negative quantities and in its creative potential can only be vaguely imagined. Modern physics, moreover, describes space as the original state of the universe right after the Big Bang, the point from which all existence originated. The same holds true for time (Davies 1983; 1995, 10-26).

Seeing the notion in a wider context not only helps to understand "space" as the first stage in the *Hengxian*, but it also illuminates just how amazing its description is. Since space and time cannot be separated, its postulation of space at the first beginning of all also means that time is at the root of the cosmos. While the *Hengxian* leaves the connection open, the *Huainanzi* clearly speaks of "space-time" in this context.

In Chinese philosophy, both space and time are real entities. Space typically indicates not only a physical location but also all the different things that inhabit it. Both by necessity must evolve and manifest in the process of cosmic unfolding, and the "domain" in the *Laozi*, described by Wang Bi as some space that "cannot be named," accordingly forms one of its stages. In the *Hengxian*, space arises from constancy, and the prime state of Constancy Before is the root stage from where space evolves quite naturally: "It fulfills itself without repressing itself."

According to the Daoist understanding of nature, the idea of "self-birthing" or "self-engendering" in the *Hengxian* can mean two things. First, it may indicate the ultimate cause of the inherent and automatic transformation and unfolding of the universe, a notion that has nothing to do with anything along the lines of "divine will." Second, it may also express one particular state of cosmic unfolding, receiving impulses from one and acting upon another. As the text says, "The birthing of constant energy does not happen in isolation."

Most important in this context is the manifestation of the inherent tendency toward unfolding: "Dark and obscure, it does not rest but seeks to give birth." The universe may be dark, obscure, and in murky confusion, but it incessantly transforms and give birth to life because it is inherently active. "It seeks, desiring to return of its own accord." In other words, *Hengxian* cosmogony is a theory of self-birthing—one of its unique structural characteristics. This idea self-birthing does not reach as far as the various beings that mutually influence and interact with each other in the world, but it demonstrates how the process of unfolding reflects the inherent nature and inherent qualities of the universe. There is no need for steering by any high god: it can perfectly move itself.

Energy

Reaching the stage of "energy," moreover, the universe moves on to reach "material existence," "beginning," and the "passage of time." Both space and constancy are formless and characterized by emptiness and nonbeing; material existence should be their opposite, characterized by having form and being. However, the text does not clarify what concrete object it indicates; left to guess, we suppose it refers to "heaven and earth." The text describes them as both, birthed from energy and the greatest objects having form. As Zhuangzi says about his wife, "In the midst of the jumble of mystery and wonder, a transformation took place and there was energy. Energy transformed, and there was form. Form transformed, and there was life—she was born" (ch. 18).

Here form emerges from energy through a process of change; this form clearly indicates heaven and earth. The *Liezi* echoes this, "One is the beginning of all form and transformation. The clear and light rose up to become heaven; the muddy and heavy sank down to become earth; the blended and harmonious energy became humanity. For this reason, heaven and earth contain their essence, and the myriad beings transform and take birth" (ch. 1). This means, material existence or "being" in the *Hengxian* most likely indicates the concrete presence of heaven and earth.

A few concrete details remain unclear. When Laozi describes the cosmic unfolding from oneness to the myriad beings, he does not use the term "energy," but then he adds, "The myriad beings carry yin and embrace yang; they blend energy to reach harmony" (ch. 42). Yin and yang here refer to types of energy, and the terms illustrate the existence and transformation of natural phenomena. This, moreover, does not begin with Laozi. Already the *Zuozhuan* 左传 speaks of the "six energies" of yin, yang, wind, rain, darkness, and light (Zhao 1). The *Guoyu* 国语 uses the expression "energy of heaven and earth" and equally refers to yin and yang, then has Bo Yangfu 伯阳父 explain how their loss of balance was responsible for the great earthquake near the three rivers ("Zhouyu").

The *Hengxian* does not divide energy according to yin and yang, but describes them in terms of clear and muddy. The *Liezi*, too, notes, "The clear and light rose up to become heaven; the muddy and heavy sank down to become earth" (ch. 1). If it was indeed compiled before the Qin, it would bear evidence for the ubiquity of this concept, supplemented further by its statement, "Essence and spirit belong to heaven; bones and joints belong to earth. Things belonging to heaven are clear and tend to disperse; things belonging to earth are muddy and tend to coagulate," referring to the inherent qualities of their energy. The *Liezi* being a later compilation would make the *Hengxian* the first documentation of this thinking, fundamental to Zhou and Qin thought and exemplified variously in the *Zhuangzi*. "It must be some dislocation of the energy of yin and yang" (ch. 6); "I take my place with heaven and earth and receive energy from yin and yang" (ch. 17); "Yin and yang are great forms of energy" (ch. 25).

Under the Han, energy is often specified as "primordial." This notion inherits both, the connection of energy to yin and yang and its division into clear and muddy as related to heaven and earth. Thus, the *Huainanzi* says, "That which was clear and yang rose up and scattered to form heaven; that which was muddy and yin sank and congealed to form became earth" (3.1; Major et al. 2010, 114). Zhang Heng 张衡, in his *Lingxian* 灵宪, similarly notes, "Thereupon, primordial energy split: hard and soft divided; clear and muddy established their positions" (Liu 1987, 3215). The *Shuowen* has, "When primordial energy first divided, clear and light yang became heaven, heavy and muddy yin became earth, and the myriad beings were arraigned between them." This clearly links the creation of heaven and earth with clear and muddy energy—a very common feature of Han thought, but in pre-Qin times found mainly in the *Hengxian*.

The two terms "clear" and "muddy" originally indicate the state of water as being either pure or murky; by extension they could also refer to the quality of sound. The *Laozi* has, "Merged and undifferentiated, it is like muddy water. Who can bring muddy water to stillness and thereby make it clear?" (ch. 15). While it uses the terms here in connection with water, it also uses them more metaphysically. "Heaven attained oneness and became clear; earth attained it and became stable" (ch. 39). Here "clear" does not refer to any kind of water or energy but indicates the quality of heaven.

Confucians, too, use the same dichotomy, applying it to human affairs and political order, moral good and evil, love and hate. As the Confucian thinker Mencius (*Mengzi* 孟子) says,

> There was a boy singing, "When the water of the Canglang is clear, it is good enough to wash the strings of my cap; when the water of the Canglang is muddy, it is good enough to wash my feet."
>
> Confucius said, "Hear what he sings, my children. When clear, then he will wash his cap-strings with it; when muddy, he will wash his feet with it. This different application is brought by the water on itself."

> A man must first despise himself, only then will others despise him. A family must first destroy itself, only then will others destroy it. A state must first smite itself, only then will others smite it. (*Mengzi* 4A8)

The *Xunzi* makes a similar point,

> The human mind may be compared to a pan of water. If you place the pan upright and do not stir the water up, the mud will sink to the bottom, and the water on top will be clear and pure enough to see your beard and eyebrows and examine the lines on your face. But if a slight wind passes over its surface, the submerged mud will be stirred up from the bottom, and the clarity and purity of the water at the top will be disturbed so that it is impossible to attain the correct impression of even the general outline of the face. (21.10; Knoblock 1994, 107)

> When the spring is clear, its water flows clear; when it is muddy, so is the water. (12.5)

All these show how the qualities of clarity and muddiness begin with water, then are extended to human and government affairs.

Heaven, moreover, is typically above, while earth is below; heaven is empty and spacious, while earth is full and solid; heaven tends to disperse, while earth tends to coagulate. The *Hengxian* is the earliest extant source to express the distinction of energy in terms of the qualities of clear and muddy, as well as the first to connect this to heaven and earth: "Muddy energy gave birth to earth; clear energy gave birth to heaven."

In Laozi's cosmogony, on the other hand, heaven and earth do not appear at such an early stage, and it is not clear where exactly they fit into the system. It says,

> The spirit of the valley does not die.
> It is called the mysterious female.
> The gate of the mysterious female
> Is called the root of heaven and earth. (ch. 6)

> There is a being, in chaos yet complete.
> It precedes heaven and earth.
> Silent it is, and solitary;
> Standing alone, it never changes.
> It moves around, yet never ends.
> It can be considered the mother of heaven and earth. (ch. 25)

According to this, heaven and earth are born directly from Dao and occupy a special position in the universe, forming two of the "four great things in the domain" (ch. 25), but they do not appear as a separate stage—unlike in the *Hengxian* and *Huangdi sijing*.

Like the *Taiyi shengshui*, where Great Oneness gives birth to water as the immediate precursor of heaven, the *Hengxian* has energy dividing into clear and muddy, giving rise to heaven and earth, from where then form and material existence arise. From here, there is universal "beginning," and the myriad beings come into existence. Thus, the text says, "Material existence emerges from space; life emerges from material existence."[5] Only in this place, it leaves out energy as intermediary.

The Process of Beginning

The text continues,

> [Beings] proliferate and give birth to each other, stretching to fill heaven and earth. They emerge from the same source but give birth to different things. Accordingly, they give birth to that which they desire. Splendid indeed are heaven and earth! They are motley and . . . manifest in numerous diverse [beings].
>
> First, there is goodness, order, and no disorder. Once there are humans, there is not-good. Disorder emerges from human beings.

After attaining material existence, full of form and complete with heaven and earth, the universe in the *Hengxian* model moves on to "beginning" and from there to the "passage of time." The former, it seems, generically indicates the first stages of all sorts of things and affairs, while the latter refers to the ongoing cycles of growth and decline, birth and return of the myriad beings. It is not clear, though, what exactly the text means by these terms. To understand them better, let us look at some comparative passages.

The *Xunzi* has, "heaven and earth are the beginning of life" (9.18), and, "The arising of all manner of beings must come from the beginning" (1.7). The *Laozi* notes, "Nonbeing is called the beginning of heaven and earth; being is called the mother of the myriad beings" (ch. 1), contrasting heaven and earth with the myriad beings as well as nonbeing with being or material existence. It also says, "The myriad beings of the world are all born from material existence; material existence arises from nonbeing" (ch. 40). Here nonbe-

---

[5] Li Ling reads "life" (*sheng* 生) as "inherent nature" (*xing* 性), and most scholars agree, but I am not so certain.

ing underlies existence and gives rise to it,[6] which matches the passage from chapter 1. In the latter, moreover, material existence is "the mother of the myriad beings," a feature that Chinese philosophy more commonly associates with heaven and earth, placing the myriad beings between them.

For example, among *Yiing* commentaries, the *Xici* states that the original nature of heaven and earth is that "they transform and give birth to the myriad beings," their "great virtue" being the ability to "give birth." The *Xugua* 序卦 says, "Once there are heaven and earth, the myriad beings take birth through them." The *Xunzi* echoes this, "What heaven and earth give birth to are the myriad beings" (10.6). They can give birth to living beings, moreover, because they possess an immense power to develop and produce all manner of things. The *Yijing* notes, "Heaven and earth give their impulse, and the myriad beings are born" (Hex. 31); and, "If heaven and earth had no interchange, the myriad beings could not flourish" (Hex. 54). The *Zhuangzi* adds to this that heaven and earth evolve and are activated in nonaction, naturally following their inherent self-nature and giving birth to life. "Heaven rests in nonaction and becomes clear; earth rests in nonaction and becomes stable. The myriad beings are born through them" (ch. 18).[7]

Based on this, we can understand what the *Hengxian* means when it speaks of life arising from material existence, the latter placing a particular emphasis on the presence and nature of heaven and earth. "Life," moreover, indicates the myriad beings taking birth and is thus their "beginning." Energy in this context is the force that fills the space between heaven and earth, their giving birth of the myriad beings constituting its concrete function. Energy, as the *Hengxian* says, is spiritual and wondrous. Once it begins to function and gets active, the host of things in the world arises. The myriad

[6] The bamboo manuscript leaves out the second *you*, so that all beings "are born from being and are born from nonbeing." This eliminates the more fundamental position of nonbeing, placing them both on the same level, which is common in the *Laozi* and quite different from Wang Bi's interpretation.

[7] Other passages include also the *Yueque shu* 越绝书, which says, "Dao brings forth energy; energy brings forth yin; yin brings forth yang; yang brings forth heaven and earth. Once they are established, there is cold and heat, dryness and dampness, the sun and the moon, the stars and the planets, the four season, and all the myriad beings" (Waizhuan 外传, "Zhenzhong" 枕中 16). The *Liezi* states, "Therefore, heaven and earth contain essence, and the myriad beings are born from them" (ch. 1). The *Lunheng* has, "Thus, human beings are born in heaven and earth, like fish are in water and maggots in human guts. They are born through energy and divide into many different species; but in that they belong between heaven and earth, they are all the same" ("Wushi" 物势). "A lord of men rules his state like heaven and earth bring forth the myriad beings" ("Feihan" 非韩).

beings all come from the energy flowing through heaven and earth, yet they are all different and unique, each with their own trajectories and expectations. The radiant presence of heaven and earth produces the most amazing richness of life, many millions of species and individual specimen.

Now, in the *Hengxian*, the birth and random division of the myriad beings into numerous species form one structure of joint emergence. There may be more or fewer species, each arising and passing away, growing and declining, transforming in never ending cycles. This, then, is the ongoing process of beginning and the continued process of the passage of time. In Genesis, God created the host of living beings in one fell swoop and made them disperse throughout the world; in the *Hengxian*, they came about through gradual emergence and unfolding in an orderly, ongoing, step-by-step process. As the text details it in a slightly different variation, "Material existence emerges from space; life emerges from material existence; sounds emerge from life; speech emerges from sounds; names emerge from speech; undertakings emerge from names." This pattern, moreover, echoes the *Heguanzi*,

> Once there is one, there is energy.
> Once there is energy, there is intention.
> Once there is intention, there are plans,
> Once there are plans, there are names.
> Once there are names, there is form,
> Once there is form, there are undertakings,
> Once there are undertakings, there are contracts.
> (ch. 5; Wells 2013, 107-08)

In other words, the *Hengxian* model of cosmic unfolding focuses on two intermediate phenomena: space between constancy and energy, on the one hand, and material existence—the emergence of heaven and earth—between energy and beginning, on the other. When the text says, "life emerges from material existence," it does in fact mean "life" and not "inherent nature," however close those two terms were in pre-Qin literature. Heaven and earth give birth to the myriad beings, and life is one aspect of this unfolding. From life, the universe moves on to "names" and "undertakings," each coming forth from the other, so that all existence is connected in an ever moving chain of mutual interrelationships. The particular sequence here, moreover, may well reflect the most visible and amazing aspects of being—life, humanity, and language.

Differentiation

The *Hengxian* next focuses on the emergence of opposites and the development of differentiation:

> First, there is the center, and only then the periphery. First, there is the small, and only then the big. First, there is the soft, and only then the hard. First, there is the round, and only then the square. First, there is the dark, and only then the light. First, there is the short, and only then the long.
>
> Once the Dao of heaven is laid out, only through oneness do things appear as one; only through return do things appear as returned.

On the horizontal plane, this means, among beings as already existing, there are mutually interdependent relationship patterns: inner and outer, small and large, soft and hard, round and square, dark and bright, short and long, and so on. Mutual opposites, they all depend on each other and transform with, through, and into one another. There is no telling which came first and which came second, which gave rise and which was born.

Laozi, too, stresses the mutual dependence and connection of interdependent opposites.

> Being and nonbeing produce each other;
> Difficult and easy complete each other;
> Long and short contrast each other;
> High and low distinguish each other;
> Sound and voice harmonize with each other;
> Front and back follow each other. (ch. 2; Chan 1963, 140)

Laozi, of course, tends to favor one side over the other within each pair of mutual opposites and encourages his readers to guard that particular tendency. As he says,

> Yield to become whole;
> Bend to become straight;
> Release to become full;
> Let go to become new;
> Lessen to attain. (ch. 22; Chan 1963, 151)

The examples could be multiplied further. The passages show Laozi's particular way of working with the ongoing cycles of the interaction of opposites. Whether the *Hengxian* echoes this or focuses more on the giving birth and being born of beings in the ongoing passage of time is hard to determine. The various examples it lists seem completely obvious at first glance, dealing with the inherent circumstances of growth and decline and reflecting a logic of gradual increase and systematic unfolding. This closely matches Laozi's words,

> A tree as big as a man's embrace grows from a tiny shoot.
> A tower of nine stories rises from a heap of earth.
> A journey of a thousand miles begins with the first step. (ch. 64)

Each of the myriad beings has its own inherent nature and unique characteristics, forming the foundation of their distinctions and differences and giving expression to their multiplicity. Their mutual relationships, moreover, are redundant and repetitive. As the *Hengxian* puts it,

> Dusky and disquiet, seeking that which gives birth to them: difference gives birth to difference, returning gives birth to returning, divergence gives birth to divergence, opposition gives birth to opposition, and dependence gives birth to dependence.
>
> [Things] seek, desiring to return to their own accord. Return is the process of life. Muddy energy gives birth to earth; clear energy gives birth to heaven. Energy is truly numinous!

In other words, things of a similar nature engender each other, following the pattern, "A gives birth to A." But what do the different "A" items signify?

According to Li Ling, and debated variously (e.g., Zhao 2005), they each reflect different feelings and emotions. He reads "caring" (*yi* 翼) for "difference" (*yi* 异), "fear" (*wei* 畏) for "returning" (*gui* 鬼), "admiration" (*wei* 愇) for "divergence" (*wei* 韦), "sympathy" (*bei* 悲) for "opposition" (*fei* 非), and "sadness" (*ai* 哀) for "dependence" (*ai* 哀).

To me, this creates more confusion than clarity. Classical emotions include joy and anger, sadness and happiness; we can say that they mutually bring each other forth—"extreme happiness creates sadness"—but we cannot claim that they reinforce themselves, as in "happiness gives birth to happiness," or "sadness gives rise to sadness." This means that the various items listed here must refer to something else, possibly concrete entities in the process of creation and transformation. Several possibilities present themselves.

One follows the model, A changes into B, as in, "The five phases give rise to each other," or, "Being and nonbeing give birth to each other." Another works with the concept that like and like enhance each other, items of the same class produce more of the same. The former is more prominent in the *Zhuangzi*, but the *Hengxian* seems to favor the latter. Other texts echo this. For example, the *Guoyu* notes, "When grasses and trees come forth, they each are their own" ("Jinyu"), "own" here meaning a specific kind or species. The *Xunzi* says, "Grasses and trees grow in abundance; birds and beasts flock in multitudes—each being follows its own species" (1.7).[8] The *Liji* similarly states, "The way the myriad beings work, they each move in accordance with their particular species" (ch. 17). There is also the proverb, "Plant a melon, get a melon; plant a bean, get a bean," roughly matching the English, "The apple doesn't fall far from the tree."

In addition, graphology suggests a connection to nature. All four characters in the passage are also written with the "grass" radical, denoting respectively a kind of hemp (*yi* 异; so defined in the *Shuowen*), a mountain weed (*wei* 萬; *Yupian*), a reed (*wei* 苇; *Guangyun*), and a vegetable (*fei* 菲; *Shuowen*). The last word, *ai* 哀, has the radical "garment" (*yi* 衣), which connects to the word "enormous" (*yin* 殷), often also written with "grass" and glossed as a type of vegetation (*Jiyun*).

This suggests that the five items listed in the *Hengxian* refer to concrete objects, things growing in nature. When it uses the formula, "A gives birth to A," in other words, it indicates that the same species of plant continues to reproduce or return to itself, emphasizing the internal integrity of species. Each type of plant has its own unique characteristics, and whatever it produces shares these, reflecting its particular abilities and tendencies. Thus, it continues the passage, "[Things] seek, desiring to return of their own accord. Return [reproduction] is the process of giving birth to life."

Creative Continuity

All living beings, including plants, continue to propagate through ongoing circulation and eternal return. Still, they each also favor certain tendencies and directions of change. If any entity ever gave birth to something completely different, it would be going against its particular "distinctiveness," its "difference," instead bringing forth a mutation. The text expresses this continuity by saying,

---

[8] Confucians also propose the concept of different species impacting each other. For example, the *Wenyan* 文言 commentary to the *Yijing* has, "The Master said, 'Those of similar sounds mate with each other, those of similar energy pursue each other. Water flows toward moisture, fire reaches for dryness, clouds follow dragons, winds follow tigers, sages act and the myriad beings rejoice. Connected to heaven, they are linked above; related to earth, they are linked below, thus each has its own species."

> If space were not space, then there would be no referring to space. If material existence were not material existence, then there would be no referring to material existence. If life were not life, then there would be no referring to life. If sound were not sound, then there would be no referring to sound. If speech were not speech, then there would be no referring to speech. If names were not names, then there would be no referring to names. If undertakings were not undertakings, then there would be no referring to undertakings.

Traditional logic teaches us that, if A is A, it cannot be not-A. The *Hengxian* expresses the same idea by saying that, if A is not-A, it cannot be called A. This also means that, if A is A, it can and should be called A. From the perspective of the inherent nature of beings, then, if any being does not match its inherent nature, it cannot be that particular being and cannot be called such. Aristotle, as described by Hegel, claims that whatever is natural to a given being constitutes its inherent nature, including also its developmental trajectory or purpose. Each being in life pursues the fulfillment of its inherent nature and transforms in accordance with its purpose (1955; 1983, 309-10). Hegel further asserts that the best possible way of being is in close accordance with its conceptual ideal. He says,

> Thus, we speak of a true friend; by which we mean a friend whose manner of conduct accords with the notion of friendship. In the same way, we speak of a true work of art. Untrue in this sense means the same as bad, or self-discordant. In this sense, a bad state is an untrue state; and evil and untruth may be said to consist in the contradiction subsisting between the function of notion and the existence of the object. (1892, 52; 1981, 86)

Hegel's argument here echoes Plato's conviction that conceptual ideals and abstract ideas are most complete, while concrete, lived reality can never quite reach them and is bound to perish. Nevertheless, Hegel also asserts that "each being as it matches its inherent nature is fully real; each being as it is fully real matches its inherent nature" (1981, 43). Looking at the issue from the perspective of the relation of names and reality, if a thing is called something but in fact is not that thing, then it cannot be said to be such. The *Hengxian* passage, "If space were not space . . ." then, deals with the relationship of names and reality or imagination and concrete presence. It emphasizes that all beings always live and act in accordance with their inherent nature. However, from the angle of concrete emergence, each being's inherent nature is in constant unfolding and transformation and, while it must evolve closely matching its inherent nature, it also has to adapt to the circumstances around it. The better the circumstances it encounters, moreover, the more it can reach its optimal state and fully realize its existence.

The Passage of Time

The *Hengxian* continues,

> In its giving birth to [things], constant energy relies on . . . returning to that which they desire.[9] Brilliantly heaven proceeds, only by return does one avoid extinction. When knowing is complete, one's far-ranging thought will not be destroyed.
>
> Material existence emerges from space; life emerges from material existence; sounds emerge from life; speech emerges from sounds; names emerge from speech; undertakings emerge from names.

The stage here called the "passage of time" indicates the level after creation, when the myriad beings live and function in the world. In the complex process of their incessant reproduction and never ending transformation, "passage of time" points to the natural law that like begets like as well as to the common fact of mutual exchange and impact. Laozi calls this "reversal" (*fan* 反) and says that it is an expression for the movement of Dao. He also uses the term "return" (*fu* 复). Both words indicate motion and point to the active reality of life, but most essentially characterize the ultimately changing nature of Dao. Laozi says:

> Attain utmost emptiness,
> Maintain steadfast stillness.
> The myriad beings jointly come to life,
> And I see thereby their return.
>
> All beings flourish,
> But each one returns to its root.
> Return to the root means stillness
> It is called return to destiny. (ch. 16; Chan 1963, 147)

Zhuangzi similarly speaks of the "multiplicity of transformations" and the "changes of energy" to describe the ongoing flow of life, saying, "The myriad beings all come from the same seed, and with their different forms they give place to one another. Beginning and end are part of a single ring and no one can comprehend its principle. This is called heavenly balance" (ch. 27). He also notes that "all beings through form give birth to each other" (ch. 22), expressing a thinking that focuses on never ceasing transformation, as Hu Shi calls it, "a theory of ongoing change" (1991, 178).

---

[9] For this phrase, see Li Rui 2004; Zeng 2006c, 160-67.

More than that, however, his thinking reflects a theory of cyclical, mutual reproduction. Zhuangzi speaks of the "mysterious workings of life" that allow each being to connect to and transform into another in active living rather than existing in isolation, in a life just its own. Life consists of beings interacting with, and changing into, the unique nature of other beings: "The myriad beings come out of the mysterious workings and go back into them again" (ch. 18).

Echoing Laozi, Zhuangzi adds ideas of reversal and return, of recovering the inherently natural. "Do not let the humanly destroy the heavenly; do not let the purposeful destroy destiny" (ch. 17). He also believes that only by returning to a primitive society can the human tendency toward regression and mutation be controlled and overcome. In both ancient thinkers, return thus takes on a twofold meaning. For one, it means the natural tendency of all living beings to return to their original state, to undergo the cycle of living and dying. For another, it also means the potential of recovery of their original nature, the ongoing process of "return to the root and recovery of perfection," what Aristotle calls the "realization of self-concordance." This is what the *Hengxian* means when it speaks of the "passage of time."

In an earlier passage, the *Hengxian* says,

> Once the Dao of heaven is laid out, only through oneness do things appear as one; only through return do things appear as returned. In its giving birth to [things], constant energy relies on . . . returning to that which they desire. Brilliantly heaven proceeds, only by returning does one avoid extinction. When knowing is complete, one's far-ranging thought will not be destroyed.

Pang Pu reads the word "Dao" in the first phrase as "earth," so it refers to the existing universe, but does not provide an argument why (2004, 22). I think he is right, and support comes from the *Huangdi sijing*, which has the phrases, "Once heaven and earth are complete," and, "Once heaven and earth are stable," matching the text here. Once heaven and earth are in existence and have taken form, "oneness" and "return" become the dominant functions.

"Oneness" here indicates the underlying unity of all beings, their common character as living entities. "Return" means their tendency to preserve and recover their inherent nature, the ongoing process of life.

The Human World

The next section of the *Hengxian* has,

> Auspicious appropriateness, beneficial craft [see Dong 2004], and diverse things emerge from basic working. When there is working, there is an undertaking. When there is no working, there is no undertaking.
>
> Regarding the undertakings of the world, they will work of themselves to become undertakings. How, then, could they not be continued?
>
> In general, when speaking names: those who come first are uncertain and speak wildly. Those who come later collate and compare what they have said. [10]
>
> Regarding the names of the world: they are emptily established. Through practice, they become inalterable.
>
> Regarding the workings of the world: the valiant bring about great workings in the world, yet they shade their radiance and appear not to be doing anything [following Liao 2003a, 90-91].
>
> As for workings, how could there be achievement and non-achievement? Neither of the two can be dispensed with.

This passage repeats the phrase, "regarding the . . . of the world," creating a rhythmic pattern in the presentation. Pang Pu considers it a strong point in the text (2004, 22-23). The passage is essential for understanding its vision of humanity: humans being part of nature, it reflects its general take on nature; humans standing in contrast to nature, it indicates just how they are similar or different. Daoism, as expressed in its classics, generally emphasizes ideas of honoring nature and acting in nonaction. It encourages human beings to align themselves or even become one with pure and perfect heaven, to give up all tendencies toward interference and control, eradicate manmade systems of rites and music, as well as all formalization and fossilization created through literature and art, norms and education.

Where, then, does the *Hengxian* stand on this? Does it share the same tendency? If we understand the text correctly, the *Hengxian* does not share Zhuangzi's vehement rejection of norms as expressed in language or "names," but rather proposes their correction, seeing them as fundamentally unavoidable in human society, reflecting a key point of difference between Laozi and Zhuangzi. It essentially makes four points:

1) Names emerge from speech as their origin.

2) Human undertakings arise through names, proceed in accordance with them, and are justified through them.

---

[10] This phrase is read differently, depending on punctuation. But no matter how the text is interpreted, it does not reject the power of speech and names.

3) Names have to match concepts and realities to be properly activated, they have to have matching connotations and work in a regular fashion.

4) Names, once formed, become a reality of their own, turning into abstract rather than concrete entities. Once people are used to them, they are hard to change, becoming fixed and stable.

All this shows that the *Hengxian* does not reject names and enhances our understanding of Huang-Lao thought in this respect. Very concretely, it expands the notion of the self-being of the hundred clans into a political doctrine of self-doing, closely matching Laozi's philosophy of self-being and its variations of self-transforming, self-balancing, self-correcting, self-aware, self-purifying, and the like (see Cao 2006a, 123-48). It says, "Regarding the actions of the world: by neither avoiding nor partaking in them, they act of themselves." This means that an enlightened king when governing the world should neither try to avoid things nor engage strongly with things but let the people "pursue their undertakings as they do." The "Daofa" of the *Huangdi sijing* echoes this, "Undertakings are neither small nor big, but just the way beings act of themselves."

Working and Acting

The last paragraph of the *Hengxian* reads,

> Regarding the actions of the world: by neither avoiding nor partaking in them, they act of themselves.
>
> The same holds true regarding life in the world. It endeavors to have nothing that does not return.
>
> Regarding the workings of the world: if they do not go against constancy [see Ding 2009, 129], there will be nothing that opposes them.
>
> Regarding the workings of the world: there are none that do not get constancy and have their results passed down. How is it that there are some who attain it and some who lose it?
>
> Regarding the names of the world: there is none that can be dispensed with.
>
> Regarding the enlightened kings, enlightened rulers, and enlightened scholars of the world: how can one seek but not consider [this]?

The *Hengxian* obviously is not opposed to undertakings, affairs, things to do in the world (*shi* 事), but they, too, should arise and unfold "of themselves." In this manner, they are manifestations of constant Dao, unalterable and beyond improvement. "Regarding the undertakings of the world, they will work of themselves and become undertakings. How, then, could they not be continued?" Enlightened kings, therefore, honor the self-acting and self-working of the people. Only then can they be valiant. As the text has it, "Regarding the workings of the world: the valiant bring about great workings in the world, yet they shade their radiance and appear not to be doing anything." This means that people of true valiance bring about great things in the world, yet they seem to be dull and never do anything. It is close to the *Laozi*,

> A good ruler brings things about and stops.
> He never dares to be actively valiant.
> He brings things about and never brags.
> He brings things about and never boasts.
> He brings things about and never takes pride.
> He brings things about but only as an unavoidable step.
> He brings things about and is never openly valiant. (ch. 30)

The context of this passage is unavoidable warfare, contrasting valiance with weakness and winner with loser. As far as Laozi is concerned, there is only one thing worth bringing about and that is certainly not massive force! However, in ordinary circumstances, a show of force may be necessary, and Laozi advises to do just that, yet without bragging, boasting, taking pride, without relying solely on one's own ability, taking possession, claiming credit, or in other ways becoming self-important (ch. 2). The *Hengxian* affirms that valiance can bring about "great workings," but only from a position of humility and withdrawal, by "shading one's radiance."

It emphasizes variously that all working and acting should honor the rules and move in accordance with constant Dao. "Regarding the workings of the world: if they do not go against constancy, there will be nothing that opposes them." Not going against constancy means never to oppose constant Dao. Along the same lines, Laozi says, "Who does not lose his place [with Dao] will endure" (ch. 33), meaning that one should not lose sight of one's role in the greater scheme of things.

Despite all this, the ultimate outcome of all working and acting is never certain. "Regarding the workings of the world: there are none that do not get constancy and have their results passed down." This matches the earlier phrase, "When knowing is complete, one's far-ranging thought will not be destroyed." The *Hengxian* maintains that any workings in the world can only be successful inasmuch as they go along with constancy. In actual life, this means that sometimes people attain it and some times they don't, some actions bring about results and some don't: "There are some who attain it and some who lose it."

Going along with constancy, even when facing the myriad beings, also means to give up all a-priori subjective knowledge. Li Ling accordingly interprets the passage, "Splendid indeed are heaven and earth! They are motley and . . . manifest in numerous diverse [beings]. First, there is goodness, order, and no disorder; once there are humans, there is not-good. Disorder emerges from human beings," to mean that in the beginning, even though there was a multitude of variegated beings in the world, goodness and order prevailed, and there was no disorder (Liao 2004, 88). Pang Pu punctuates slightly differently, reading the passage to mean that goodness prevailed and governmental order kept disorder at bay (2004, 22). This seems plausible, as is the addition of "beings" after "diverse."

On the other hand, "diverse" (*cai* 采) appears variously as a noun in ancient texts. For example, the *Zuozhuan* speaks of the "six diversities," indicting six colors (Zhao 25). The *Hengxian* would accordingly indicate the multihued variety of living beings. The *Hanfeizi* says, "Therefore, the sage holds on to oneness in stillness, lets undertakings follow their own destiny and settle themselves. Disregarding diversity, he maintains plainness" (8.2). Here diversity is the opposite of plainness, indicating cultural sophistication.

The text further notes that "at first, there is goodness," contrasting the state of "at first" with the time when humans existed. According to Pang Pu, this first state is a time "before" the myriad beings came to be, a primordial level at the very beginning (2004, 22; Cao 2006, 120). Moral characteristics, that is to say, depend on the sheer existence of living beings, goodness being part of primordiality, and the rise of humanity bringing about evil. This understanding goes radically beyond Zhuangzi's vision of a primitive society.

The *Hengxian* clearly contrasts a state before beings with the existence of humanity, goodness with not-good, and order with disorder. The three pairs are further linked in a causal relationship: because it is before beings, there is goodness, leading to a state of order; because humanity exists, there is not-good, resulting in disorder. This thinking is quite close to concepts of self-so and nonaction typical for Laozi and Huang-Lao. Thus, the *Guanzi* says, "Don't place humans before all beings, but observe their patterns. If, once you act, you lose your place, rest in stillness, and you will attain it by yourself. Dao is never far off, yet difficult to fulfill." It also has, "Still yourself and wait on things; when they come to you, names will follow of themselves;" "Being in accordance with beings is like partnering them;" and "Shade yourself and made beings your law" ("Xinshu shang"). The *Hanfeizi* similarly says, "If principle shifts before beings do anything, we speak of foreknowledge. Foreknowledge means having strange impressions without clear cause" (ch. 20).[11]

Based on this evidence, "before beings" means acting in accordance with beings, following the objective order and rules of the universe and avoiding all subjective, personal judgment and assertive action, thus the admonition, "Don't act before beings." The existence of humanity, then, means the presence of intentionally structured governmental order; it is the watershed between good and evil. The enlightened ruler works hard to act in due accordance with the self-existing (natural) laws and patterns while avoiding all subjective thinking and opinions. This closely echoes the *Hengxian* description of actions in the world as "not going against constancy" and maintaining a level of "knowing that is complete."

Laozi's political philosophy, too, contrasts assertive action with nonaction as ways of governmental administration and strongly opposes any form of intervention on the side of the ruler. Instead, he should go along with the needs and desires of the people in nonaction. He describes the realities of self-serving, interventional government quite vividly,

> The courts are exceedingly splendid,
> While the feeds are covered in weeds.
> The granaries are depressingly empty,
> While they wear elegant robes.
> They carry sharp weapons
> And relish grand banquets,
> Having wealth and treasures accumulated in excess.
> This is robbery and extravagance.
> Not at all Dao! (ch. 53)

The *Hengxian* uses the contrast of "working" and "non-working" as well as "undertaking" and "non-undertaking" to flush out the difference between assertive action and nonaction in concrete terms, manifesting in "auspicious appropriateness, beneficial craft, and diverse things." Its ultimate conclusion is that "enlightened kings, enlightened rulers, and enlightened scholars" should maintain an attitude of "non-undertaking" and "non-working" in the hope that this will bring about ideal government.

---

[11] Similar ideas also appear in the *Huainanzi*: "What we mean by 'nonaction' is not acting before beings" (ch. 1).

Conclusion

From the first arising and unfolding of the universe to the emergence and activation of the myriad beings, and again to the order of human society, the rather short *Hengxian* manuscript presents a cosmology and anthropology of vast richness. In many ways, its thought reflects Daoist concepts and visions, and scholars have typically placed it within the lineage of early Daoism. The question is, though, where exactly should it be located in the overall unfolding and development of the school? My suggestion is to see it between Laozi and Zhuangzi and to use the simultaneously discovered *Taiyi shengshui* to get a better sense of the state of Daoist thought at the time, eliminating some of the gray zones in its history. Of course, this inquiry presents only a first step, which in many ways follows the path-breaking work of Li Xueqin.

We can conclude that the *Hengxian* comes after the Laozi for two reasons. One, it stands in the tradition of Daoism founded by Laozi; two, it is contemporaneous with the Guodian *Laozi*, which is neither the first nor the original version of the text. While the *Hengxian* does not make use of Laozi's central concepts of Dao and virtue, it has quite a few central concepts in common with the text: constancy, simplicity, stillness, emptiness, energy, space, oneness, return, self-generation, names, and more. In many ways, it inherits the overall philosophical thrust of the *Laozi*, yet it also develops its thought, proposing unique concepts that are yet based on Laozi, such as Constancy Before, Great Simplicity, Great Stillness, Great Emptiness, heavenly workings, muddy and clear energy, self-working, self-acting, and the like.

More specifically, it adopts the *Laozi* notions of simplicity, stillness, and emptiness, then expands them in its own way, making this by adding the prefix "great" to each of them. It uses the idea of constancy, already found in *Laozi* 1, using the word *heng* rather than the transmitted *chang*. In chapters 4 and 25, Laozi describes Dao as "before images and gods" or as "born before heaven and earth," using the word "before" to indicate a very long, infinite, cosmic time span. Adapting the two terms "constancy" and "before" from the *Laozi*, the text creates its own unique concept of Constancy Before to express the starting point and original state of the universe. All this strongly indicates that the text came after the *Laozi* and inherited and developed its ideas.

Who in actuality wrote the text, remains shrouded in mystery. Li Xueqin, suggests it may have been Guanyinzi who, according to the *Zhuangzi*, "expounded [Laozi's ideas] in terms of constant nonbeing and being, and headed his doctrine with the concept of Great Oneness" (ch. 33). However, the tradition claims Laozi had many disciples, including also Wenzi, Kangcangzi 康藏字, Liezi, and Yang Zhu 杨朱, so that it is quite impossible to tell who among them might be responsible for the *Hengxian*. Still, we may be able to narrow down its place in the lineage. Zhuangzi is commonly thought to have lived about 360-280 BCE—Qian Mu dates him to 365-290—the time when the Shangbo tomb was closed. Since the text must have existed before its copy was placed in the tomb, it must precede Zhuangzi.

There are, moreover, certain key differences in their thought. The *Hengxian* emphasizes the inherent stability of beings, "Difference gives birth to difference; returning gives birth to returning . . ." and "If life were not life, then there would be no referring to life." Zhuangzi, on the other hand, focuses centrally on the "transformation of energy," the ongoing change and fluidity of all existence, the mutual interdependence of things, arguing in opposition to the dialectician Huizi. His thinking is thus quite different from the view expressed in the *Hengxian*. Also, Zhuangzi is strongly concerned with mental and spiritual transcendence, while the *Hengxian* focuses on social order, working with Laozi's concepts of nonaction and self-generation, but not yet linking them to ideas of the law—which comes later, in the Huang-Lao school.

All this, then, allows us to place the *Hengxian* between Laozi and Huang-Lao (arguments of which also appear in the *Zhuangzi*). It is comparatively early, representing a different stage from that documented in the *Huangdi sijing*—which, too, documents a deep inquiry into the origins and stages of the universe. The *Hengxian*, then, is Daoist cosmology at its rawest, an important milestone in the growth of Daoist cosmological thought.

Chapter Two

Taiyi shengshui

Textual Structure and Conceptual Layers[1]

The *Taiyi shengshui*, lit. "Great Oneness Gives Birth to Water," unearthed at Guodian has elicited quite a bit of disagreement in the scholarly community, raising issues that require further examination.[2] Here I would like to address some of these issues, focusing particularly on the internal division of the text, its parts and sections, and evaluating its content in terms of philosophical concepts and theoretical understanding.

With regard to the internal division, the key question is whether it consists of one single entity or divides into two major parts. Another issue is whether or not it is part of the *Laozi* in terms of conceptual and theoretical outlook. This includes issues of whether its notion of Great Oneness is primarily a philosophical or religious concept; how it relates to Dao; what the connection is between Dao, heaven, and earth; and—more generally—how much these ideas and concepts share with Laozi, Daoism, yin-yang cosmology, and numerology (see Zhang 2012).

In addressing these various issues, I shall begin by focusing on the textual structure of the text, then address its conceptual layers, and finally examine its relationship to other early philosophical schools and works. In all cases, I base my studies on previous research, developing it into new dimensions and ways of understanding.

Overall Structure

With regard to the textual structure of the *Taiyi shengshui*, the central question is whether it essentially consists of one or two major parts, followed

---

[1] This chapter is the result of research done under grants by the National Foundation for Social Sciences and the Bejing City Foundation for Social Studies.

[2] For interpretations and discussions of the text published to date, see: Cao 2014; Chen 2000; Cui 1998; Li 2002; Liu 1999; Qiu 2004; Xing 2005; Zhao 1999. For textual studies, see Ding 2002; Tan 2007.

closely by the issue of its division into lesser sections. Most scholars follow the original editor and assume that the text consists of just one body, which they then divide into sections of various lengths. Others doubt this understanding, suspecting that the text divides into two separate parts (e.g., Ding 2002, 234).

Cao Feng is even more radical, providing a completely new reading by seeing Part Two as an independent document, which by implication also means that Part One is a separate text—obvious already from the subtitle of his study, "Part Two of the *Taiyi Shengshui* Is an Independent Work" (2014). For our purposes here, it is therefore essential that we first clarify whether or not the text consists of one or two documents, which in turn will have a major impact on our reading of its worldview and concepts.

As regards the division of the text, I have always maintained and still believe that it is one integrated document and not two. I have presented my arguments in this regard earlier (Wang 2012), and have nothing fundamentally new to add here. To avoid judgment based on preconceptions, in this discussion I begin with the actual physical bamboo slips, their arrangement and numbering, rather than their physical format and contents of their writings.[3]

Specifically, the bamboo slips contain markers for sentences (short lines) as well as for repeated and combined phrases (double lines). In addition, #14 also has a section number marking at the bottom that looks like a black nail, followed by an empty space. Based on these two pieces of information, Li Ling reached the positive conclusion that they indicate the end of a major section. He asserts that #14 marks a major division of the text, yet excludes the possibility that Part Two is an independent document (2002, 33-34).

In concrete detail, about eight centimeters of bamboo at the lower end of #14 are broken off, fortunately leaving the black-nail marking visible. After this, there is an empty space, worth about the length of three characters. This allows the conclusion that the missing eight centimeters were originally without writing. The physical format, writing style, and overall arrangement are also confirmed by the Guodian C version of the *Laozi.* This consists of four sections connected in their writings that show empty spaces after section numbers. All this suggests that the black-nail marking on #14 signifies a division in the text; it does not necessarily imply that Part Two is a separate document.

---

[3] Judging from the contents and meaning of the text whether the *Taiyi shengshui* consists of one document or two allows arguments in all different directions, therefore, it is best not to apply this particular method.

Regrettably, the other bamboo slips of the *Taiyi shengshui* do not have any section markings or numberings. Li Ling, who concludes that #14 is the end of the one and only section of the text, is unwilling to face this reality and speculates about the fact that other slips do not have markings. "Precisely speaking," he says, "it is the case that we have not discovered any section markings or numberings mainly because the lower part of #8 is missing. If we reversed the order of the two parts, it is not impossible that there might be section markings at the end" (2002, 32). Basing his evaluation on the markings of #14 and #8, Li Ling seeks to prove that the *Taiyi shengshui* "is a single document." When he says that the bamboo slips can be read consecutively, that is what he means.

In this point, he is in complete agreement with Qiu Xigui. In the latter's judgment, besides the missing parts on excavated bamboo slips there cannot be any lacunae. He says, "The Guodian C version of the *Laozi*, found together with the *Taiyi shengshui*, may lack certain chapters but, in terms of content, there are no gaps. The *Taiyi shengshui*, therefore, cannot have any missing parts either" (2004, 243). The fact is, at this point we have no reliable evidence for missing bamboo slips.

All this means, then, that the black-nail markings on #14 indicate a break in the flow of the text, making Part Two a new section but not an independent document. This is also borne out by comparing it with the format of excavated Confucian bamboo materials, which come in three kinds. The first includes texts divided into four parts with particular markings, some of which further contain two sections that are also marked. They include the *Tang Yu zhi dao* 唐虞之道 (with thick, short lines as markers for both parts and sections), the *Liude* 六德(with hook-shaped dividers for parts and thick lines for chapters), as well as the *Xing zi ming chu* 性自命出 and the *Chengzhi wenzhi* 成之闻之(both with hook-shaped dividers and no section markers).

The second kind has no markings for parts but some for sections, including hook-shaped dividers. They appear in the *Liuyi* 缁衣, *Wuxing* 五行, *Laoda yishi* 穷达以时, and *Lu Mugong wen Zisi* 鲁穆公问子思, the latter showing coarse, short lines. The third type, finally, has no markings whatsoever, as apparent in the *Zunde yi* 尊德义 and *Zhongxin zhi dao* 忠信之道. The *Taiyi shengshui* matches the second kind. All its markings signal part divisions; it has no markers for sections.

In addition, based on the format of the bamboo slips and the writing style of the characters, the *Liuyi* and the *Wuxing* can be combined into one volume, as can the *Lu Mugong wen Zisi* and the *Laoda yishi*. They each show markings after the last section and, even though this signals a conclusion in terms of meaning, the markings are the same as those on earlier slips and therefore do not indicate that the section in question is an independent document. Based on this, Li Ling's view that the truncated slip #8 originally also contained end markings is not a forgone conclusion, and #14 with its markings can just as well signify a minor hiatus in the flow of the text, especially if we assume that #8 did indeed show similar signs.

Also, the original editor of the text assumed, on the basis of its closeness to Guodian C, that the two texts formed two sections of one document. He did not rely on the particular markings on the bamboo slips, but rather focused on their format, writing style (one of the main reasons why he grouped it with Guodian C), and content. This arrangement was not the result of an organized, forethought plan, but came about after careful selection and evaluation of the materials. In fact, the format of the bamboo slips is not an "absolute standard" in the decision of whether or not the *Taiyi shengshui* consists of one part or two, nor is the analysis of its content (actually, quite the contrary).

Disregarding the style and format of the physical slips, the content-based argument that the text is one unit is weak indeed, its philosophical statements suggesting that we are dealing with two separate parts or independent treatises.[4] As a result, we can save ourselves the trouble of arguing either way on the basis of content and pursue a different approach. Thus, I focus on the section markings on the actual slips in the assumption that this method has the stronger persuasive power.

Parts and Sections

Assuming that the *Taiyi shengshui* is one integrated document and does not consist of two separate parts, I would now like to turn to the question of its section division and arrangement. As well known, scholars tend to agree that the text from slip #1 to #8 constitutes one part in a clear, systematic order of presentation. The questionable division is within Part Two, the main issue being the placement of #9. Based on the most obvious division of the material, there are three key places where it could go: one is as at the beginning of Part Two; the second is between #12 and #13; and the third is before #14 (as

[4] Cao Feng criticizes the original editor for having an "organized, forethought plan" and presenting the text as one unit based on its theoretical outlook. However, he himself uses the contents and philosophical discussion of the text to argue in favor of two separate parts.

proposed by Qiu Xigui and adopted by Liu Zhao). The first two create a division that leaves the entire Part Two as one unit, while the third divides the latter into two further sections. That is to say, it creates one section from #10 through #13, called "Names and Appellations," and a second section in the combination of #9 and #14, called "The Dao of Heaven Values Weakness" (Qiu 2014, 247).

First of all, I do not agree with the third option, i.e., Qiu Xigui's way of dividing Part Two into two sections and placing #9 rather late in the sequence. This is because Part Two only has one slip with end markings and not two and because it definitely has no lost slips. Also, this division upsets the original order. Overall, I think it best that we accept the established order and forego imposing any structure based on our particular reading of the content. Naturally, as Qiu Xigui emphasizes in his conclusion, he cannot ultimately deny the possibility that what he divides into two or three chapters may in fact be just one.

Highly similar to Qiu Xigui's methodology, Ding Sixin leans toward dividing Part Two into two separate sections. He differs in that he sees one section beginning with #9, moving on into a portion he believes existed originally but is now lost, and another section from #10 to #14. He calls it "The Names and Appellations of Heaven and Earth" (2002, 234). However, this way, too, ignores the chapter markings on the actual slips and is essentially an analysis based entirely on his particular understanding and interpretation of the content. If we are serious about finding out what the original section division was and want to move beyond the present, newly imposed division, creating section titles from our imagination, we must work with the actual shape and format of the slips, honoring them as they are.

Based on this discussion, I strongly submit that Part Two consists of only of one section. However, within that understanding the question still remains where best to place #9. Continuing along the lines proposed by Qiu Xigui, I think it should come between #12 and #13. This means, the text would now read:

> [12] The names and appellations of heaven and earth stand side by side. Therefore, they exceed their areas, not allowing to match each other. . . [9] The Dao of heaven values weakness. It reduces its completion, thus adding to life by cutting back on strength and relying on [weakness. For this reason, heaven is not sufficient] [13] in the northwest. What is below it is high and strong. earth is not sufficient in the southeast. What lies above it is [low and weak]. When the insufficiency lies above, there is excess below. [14] When it lies below, there is excess above. (Chen 2002, 32-33)

This arrangement is identical with Qiu's earlier line-up, fundamentally arrived at on the basis of meaning and content. However, even from this angle a yet different reading is entirely possible. This would begin with #10 and go to #12, creating a sequence that focuses on acting in accordance with heaven and earth. It says,

> [10] Below is soil; it is called earth. Above is vital energy; it is called heaven. Dao is also their appellation; gray and murky are their names. [11] He who follows things by means of Dao must depend on the name. Therefore, things are accomplished and his life expectancy is prolonged. When the sage does things, he also depends on its name. [12] Therefore, his achievements are made and his body suffers no harm. The names and appellations of heaven and earth stand side by side.

In this case, #9 could be placed earlier, before the presentation of the best way to act in the world. This, of course, is just one possibility of dealing with the text, not unlike the various other options noted earlier, such as Li Ling's contention that the section beginning with "Below is soil . . ." should be considered the first line of Part Two. Understanding the work in this manner, the discussion of heaven and earth follows naturally.

The main doubt scholars have voiced about the second reading has to do with the question of which three characters to supply after the phrase "not intending to match each other" in #12. This is, in fact, a highly critical and unavoidable issue. However, in order to select suitable and appropriate words for the phrase, we must first get a thorough and accurate understanding of the preceding phrase, "The names and appellations of heaven and earth stand side by side. Therefore, they shift their areas." Li Ling reads the word after "therefore" as "exceed" (*guo* 过) and interprets it to mean "false" (*e* 讹) (2002, 33), while Zhao Jianwei reads it "act" (*wei* 为) and understands "intend" (*si* 思) to mean "virtue" (*de* 德) (1999, 391).

The text reads, "Below is soil; it is called earth. Above is vital energy; it is called heaven." This may well be the intended meaning of the phrase, "The names and appellations of heaven and earth stand side by side," since here the text lines up and juxtaposes the major characteristics of heaven and earth. Whatever comes after "therefore," then, can either follow this reasoning or present an opposing view. The key is how best to read 𢘍, the original character. There are other instances in the bamboo texts where it stands for "exceed," but none where it means either "false" or "act." Ding Sixin reads it to mean "lose" (*shi* 失), which is quite appropriate.

The word, after all, consists of the character for "heart" (*xin* 心) combined with the word for "change" (*hua* 化), and the fundamental meaning of "change" involves "deviation" or "loss of regularity." For example, the *Guoyu* says, "Now, the vital energy of heaven and earth never loses its rhythm. If they ever exceeded their rhythm, people would suffer from disorder" ("Zhouyu shang"). Wei Zhao 韦昭 comments, "Exceed means lose." The *Yijing* has, "heaven and earth move in order, therefore the sun and the moon do not exceed and the four seasons do not fail." Yu Fan 虞翻 interprets, "Exceed means lose the right measure." The *Guanzi* states, "You may exceed the lord, but never lose sight of the small man" ("Lizheng"). Here, too, "exceed" really means "lose."

Next, the word for "areas" (*fang* 方) originally refers to boats moving parallel or being properly lined up. The *Shuowen* clearly defines it as such. A derived meaning of the term is "even" or "aligned," from which comes "position," "direction," "standard," and "measure." Since the *Taiyi shengshui* notes that "heaven and earth stand side by side," the meaning of "aligned" or "matching" seems most appropriate. "Exceed their areas," therefore, means that heaven and earth lose their inherent balance and alignment.

The following phrase, "not intending to match each other" should really be "not allowing," replacing *si* 思 with *shi* 使, a common pattern in the silk manuscripts. The word for "match" (*dang* 当) is entirely appropriate and noncontroversial. All this leads to the understanding of the phrase as meaning that heaven and earth do not allow their different aspects to be balanced or evenly matched (Qiu 2004, 252).

Based on this interpretation, the three characters that should follow the phrase should be "and not lasting forever" (*wu jiuchang* 无久长). This echoes the *Laozi*:

> Heaven is eternal and earth everlasting.
> The reason why heaven and earth can be eternal and everlasting
> Is that they do not life for themselves.
> Therefore, they can be eternal and everlasting. (ch. 7)

The fundamental meaning is that heaven and earth are not selfish or self-oriented, which gives them the ability to endure permanently. On the other hand, if they "exceed their positions," they become self-centered and this will "not allow them to match each other," leading to a loss of the ability to go on forever.

These three characters make a lot of sense and complete the section when supplemented to #9, beginning with, "The Dao of heaven values weakness," and linking to #13. The phrase "cutting back on strength and relying on. . . ," forms the last part of #9. Both Chen Wei and Liu Xinfang read "rely on" (*ze* 责) as "accumulate" (*ji* 积) and replace the missing character with "weak" (*ruo* 弱), reading the phrase "accumulating weakness," highly appropriate both in interpretation and choice of supplementation (Chen 2000, 229; Liu 1999, 78).

Li Xiaoyu, too, acknowledges that "rely on" should be read as "produce" (2003, 144), citing three examples from the literature. One is from the *Guiguzi* 鬼谷子:"For this reason, the strong accumulates through the weak; the excessive accumulates through the insufficient: this is how to practice the arts of Dao" ("Mou" 谋). The second is from the *Liezi*,

> If your aim is to be hard, you must guard it by being soft.
> If your aim is to be strong, you must preserve it by being weak.
> What begins soft and accumulates must become hard.
> What begins weak and accumulates must become strong.
> Watch them accumulate, and you will know
> Where good fortune and disaster come from. (ch. 2; Graham 1960, 53)

The third is from the *Huainanzi*, which contains the same passage as the *Liezi* (ch. 1).

The Text

All this goes to show that the *Taiyi shengshui* is really one text, divided into two parts, and does not consists of two separate, independent documents. They can be classified either as Upper and Lower or Part One and Part Two. Here is the complete text (see also Henricks 2000, 123-29):

PART ONE

Great Oneness gives birth to water.
Water returns and assists Great Oneness, thereby forming heaven.
Heaven returns and assists Great Oneness, thereby forming earth.
heaven and earth [repeatedly assist each other], thereby forming spirit and light [*shenming* 神明].
Spirit and light repeatedly assist each other, thereby forming yin and yang.
Yin and yang repeatedly assist each other, thereby forming the four seasons.
The four seasons repeatedly assist each other, thereby forming cold and hot.
Cold and hot repeatedly assist each other, thereby forming wet and dry.
Wet and dry repeatedly assist each other, thereby forming the cycle of the years.
Thus, the process is complete.

The cycle of the years is what wet and dry give birth to.
Wet and dry are what cold and hot give birth to.
Cold and hot are what the four seasons give birth to.
The four seasons are what yin and yang give birth to.
Yin and yang are what spirit and light give birth to.
Spirit and light are what heaven and earth give birth to.
heaven and earth are what Great Oneness give birth to.

Therefore,
Great Oneness stores in water and acts in the seasons.
Revolving and again [beginning],
It takes itself as the mother of the myriad beings.
Waning and waxing in alternation,
It takes itself as the guideline of the myriad beings.
This is
What heaven cannot kill,
What earth cannot bury,
What yin and yang cannot form.
If a ruler knows this, we call him [enlightened]. [5]

PART TWO

Below is soil; it is called earth.
Above is vital energy; it is called heaven.
Dao is also their appellation;
Gray and murky are their names.
He who follows things by means of Dao must depend on its name.

Therefore,
Things are accomplished and his life expectancy is prolonged.
When the sage does things, he also depends on its name.
Therefore,
His achievements are made and his body suffers no harm.
The names and appellations of heaven and earth stand side by side.
Therefore,
They exceed their areas, not allowing to match each other
And losing their ability to last forever.

The Dao of heaven values weakness.
It reduces its completion,

---

[5] Instead of "enlightened," the word might also be "sagely," but they rhyme structure of the original suggests the former. Both Confucians and Daoists emphasized the quality of enlightenment in the ruler, exemplified in Laozi's phrase, "Knowing constancy means being enlightened" (ch. 16).

Adding to life by cutting back on strength and relying on [weakness.
For this reason,
Heaven is not sufficient] in the northwest.
What is below it is high and strong.
Earth is not sufficient in the southeast.
What lies above it is [low and weak]. [6]
When the insufficiency lies above, there is excess below.
When it lies below, there is excess above.

The *Laozi* Link

The questions that arise next are: Who was the original author of the text? And, what is its relationship to the Guodian *Laozi*? As regards the first, on the basis of the evidence available to date, Li Xueqin's (1999) speculation that it goes back to Guanyinzi is the most convincing. There is no need to repeat his arguments here. Instead, we can focus on the second question.

It is highly probably that the *Taiyi shengshui* was copied in the same volume or cluster of slips as the Guodian C version of the *Laozi*, leading to various speculations with regard to the relationship between the two texts. That is to say, it is either a separate text independent of the *Laozi* or forms part of it. Most scholars, based on the contents of the *Taiyi shengshui* and the transmitted edition of the *Laozi*, conclude that it is independent, an entirely separate document, even if it contains materials relevant to Laozi's thought.

However, there is also the understanding that it forms part of the Guodian *Laozi*. For example, Cui Renyi incorporates it directly into the bamboo *Laozi* (first group) (1998, 37). Similarly, Wen Xing believes that it was not copied together with the *Laozi* by accident, but that the two form one "coherent document" in content and structure, and that the *Taiyi shengshui* essentially constitutes part of Guodian C (2005, 247). As noted earlier, I personally still think that it is an independent work and does not form part of the *Laozi*. The two being copied in one volume does not necessarily mean that they form one document; the mere copying pattern is not sufficient to indicate sameness of text. Along the same lines, the fact that many different Confucian works found at Guodian are copied and bundled together does not mean that they form part of the same document. Conversely, the different versions of the *Laozi* do not share the same volume, A and B being in one, and C in another bundle of slips, yet they are still the same text.

---

[6] This supplement follows Cui 1998, 37.

The more fundamental problem behind this issue is the nature of the *Laozi* as text, whether the transmitted edition, the silk manuscripts, or the Han version represent the first original. Considering the *Taiyi shengshui* as part of the Guodian *Laozi* implies a disconnect between the later text in 5,000 words and the Guodian version. Alternatively, it opens the assumption that the original *Laozi* consisted only of what remains at Guodian and all other parts are later additions. However, this assumption does not hold, primarily because the chapter division of the Guodian *Laozi* is very close to that of the present edition as well as that of the silk manuscripts and the Han text. This indicates that the Guodian and the present *Laozi* form one unit, inheriting the same strand of materials.

Another reason why the *Taiyi shengshui* must be separate it is that part of the Guodian *Laozi* is a relatively complete part of the very first original version. The fact that it is so short does not necessarily mean that it is only an excerpt. It is also possible that this *Laozi* transcript forms part of a larger "buried edition," but this we cannot verify at present. What we can verify is that the earliest *Laozi* we have is basically the same as the transmitted, silk, and Han editions, which is further substantiated by its citations and references in early pre-Qin literature.

Liu Xiang's *Shuoyuan* records that Shu Xiang, a contemporary of Confucius who served under Duke Ping of Jin once cited Laozi's notion of "hard and soft," showing that the text was available from an early age. The section he refers to, moreover, today appears in two separate chapters, i.e., 43 and 76, but is not contained in the Guodian version. Chen Guying and Ding Sixin further show that quite a few *Laozi* passages cited in the *Zhuangzi* do not appear in Guodian (Ding 2002, 87-89; see also Yan 1983; Chen 1992; 2015). The view that the *Laozi* is a collection based on the *Zhuangzi* and later than the 3rd century BCE is, therefore, erroneous.

Incidentally, scholars tend to assert that the pre-Qin classics did not come about all at once and as the work of a single author, but that they evolved over many years in an extensive process of evolution. What this means precisely, I think, is not so much that a text was copied again and again, inviting changes and errors in transcription, but that its content evolved by passing through a series of editors. If this is the case, the argument creates a serious issue, since it prevents us from using any ancient text to study the philosophy of a particular thinker.

In fact, the formation of ancient texts is more complex than this understanding suggests, and we have to distinguish between several different situations. Some texts probably did not arise at a single time and place, such as the *Shangshu* 尚书, *Shijing*, and *Yijing*, which contain complex records and technical information. However, works that form the basis of the study of pre-Qin philosophers are not like this, and certainly not the *Laozi*. In the period before the Qin, individual thinking rose to the forefront and a hundred thinkers flourished, recorded in many different text, such as the *Lunyu* 论语, *Mengzi*, *Xunzi*, and even the *Huangdi sijing*. Reflecting the thought of particular philosophers and lineages, we cannot simply say that their contents was not formed at one time and place.

Conceptual Dimensions

The ongoing process of interpretation of the original classics, their transmitted style of elucidation and their particular expression of worldview and thought have generally been considered a distinctive feature of the evolution of Chinese philosophy. One effect of this feature is that some philosophers' thought does not present an integrated structure or systematic system. However, we cannot say that these philosophers' thought essentially has no structure or system, or any overall themes or arguments that give expression to their particular philosophy. A case in point is the philosophy of Mozi, Xunzi, and others, who all have very distinctive concepts and visions of reality. Seen in this light, the *Taiyi shengshui* is not a commentary on the *Laozi*, but an independent text. This means that it not only has a unique formal structure but also a practical system, a hierarchy of concepts, and its own particular logic. It presents a discussion of the universe and a fundamentally philosophical understanding of the myriad beings. It establishes a particular cosmology and worldview, quite naturally, I believe, offering an early version of Daoist philosophy.

In terms of its cosmology and worldview, the *Taiyi shengshui* elucidates the notion of Great Oneness as the origin and root cause of both the world and the myriad beings. There is no disagreement about this whatsoever. Issues arise with regard to the understanding of Great Oneness as a philosophical concept, its reading as a religious or theological notion, and its position in relationship to Dao. Generally speaking, the *Taiyi shengshui* consists of two parts, presenting a cosmogony (the origin of the world and the myriad beings) as well as a theory of the nature of things (the unified quality the world). In both parts, its highest concept is Great Oneness, forming the root cause of the universe, from which a comparatively clear series of evolutionary stages unfolds: Great Oneness← → water ← →heaven ← → earth→ spirit and light→ yin and yang → the four seasons→ cold and hot→ dry and damp→-the cycle of the years.

The text then describes the series in reverse, beginning from the later stages and showing how each later level grew from an earlier one until it goes all the way back to the origin, that is, the level of Great Oneness. Humanity generally has the urge to explain the origin of the universe in different ways, bringing forth myth and religion as well as philosophy and science. In early human thinking, these methods were not entirely separate and evolved in close connection. Later philosophical interpretations arose that differed from religion and mythology.

Great Oneness in the *Taiyi shengshui* is not a divine entity or god, nor does it precede the complex philosophical concept. In the mythical religion of the Three Dynasties, the central deity was first the thearch and then heaven, both seen as creator and ruler of the myriad beings (as evident in the *Shangshu* and *Shijing*). No other nature divinity was ever venerated as the highest god, nor were human ancestors. In the development of Chinese philosophy from the Three Dynasties to the Eastern Zhou, Confucianism most strongly emphasized inherent nature and self-awareness, while yet continuing the traditional belief in the thearch and heaven. Daoism at the same time focused more pointedly on the natural way of heaven, closely related to astronomy, astrology, and the calendar. Pursuing this line, its thinkers envisioned a supreme force, a universal and ubiquitous power to explain the universe and the myriad beings. In this process, they came to supersede earlier notions of transcendent deities, such as the thearch and heaven—both described as arising only after Dao.

At the same time, Laozi for the first time used the common numeral "one" and made it into the significant philosophical concept of oneness. Doing so, he placed it on even footing with the concept of Dao at the center of creation and all existence. Beyond these two, he also emphasized the concept of greatness. In the philosophical development after Laozi, thinkers inherited the concept of Dao as the most fundamental power as well as that of oneness. They further combined the latter with the notion of greatness and established oneness as a new fundamental dimension. Oneness as well as Great Oneness in due course became synonymous with Dao and evolved into central Daoist concepts. Already the *Zhuangzi* notes that oneness as a concept originated with Lao Dan and was central to the thought of Guanyinzi: "He dominantly focused on Great Oneness" (ch. 33). This clearly states that the concept was established in the *Laozi* and was on even footing with Dao, but particularly developed by Guanyinzi. Oneness being as much a highest concept in this cosmology as Dao, they two naturally were placed on the same level and expressed with the same terminology.

This feature is clearly visible in the *Fanwu liuxing*, the *Zhuangzi*, and the *Huangdi sijing*. Withinthe latter, the *Shida jing* in particular has Huangdi 黄帝, the Yellow Emperor, speak to Li Mo 力黑, asking whether oneness has content, to which Li Mo replies, "Oneness being the root of Dao, how can it not have content?" The the Yellow Emperor asks again, "Is oneness just one and that is all, or does it also grow?" To which Li Mo replies: "Oneness is the root of Dao. How could it be that it would not grow?" (ch. 9). [7]

Based on this, it makes sense that Qian Baocong considers oneness as the primary concept of Daoism. It also closely relates to Taiyi, the Great One, as a deity and stellar constellation, both present in pre-Qin times although, as Li Ling notes, they are not formally described in the literature until the Han. In other words, oneness even before the Qin had three different meanings and dimensions (Li 2000, 237), but I think that the deity and the constellation developed after to the philosophical concept. Obviously, the *Taiyi shengshui* is a philosophical text; it is not a fairy tale or creation myth. Oneness here is not the derivative of a deity or constellation: it is first and foremost an abstract cosmological idea.

In the *Taiyi shengshui*, Dao refers to oneness or to heaven and earth, as indicated by the word "their" or "its" (*qi* 其) in Part Two. The text has,

> Below is soil; it is called earth.
> Above is vital energy; it is called heaven.
> Dao is also their appellation;
> Gray and murky are their names.
> He who follows things by means of Dao must depend on its name.
>
> Therefore, things are accomplished and his life expectancy is prolonged.
> When the sage does things, he also depends on its name.
> Therefore, his achievements are made and his body suffers no harm.

Chen Wei, Zhao Jianwei, and others think that *qi* refers back to oneness (Chen 1999, 389; Zhao 1999; Qiu 2004, 249), but quite a few scholars also believe that it indicates heaven and earth (Li 2002, 39). The sentence before "Dao" speaks of heaven and earth; the phrase following the passage notes that "the names and appellations of heaven and earth stand side by side." In addition, the text says that "Dao is their appellation," making it easy to see Dao from the perspective of its relationship to heaven and earth.

---

[7] The *Daoyuan* similarly says, "Now be oneness and do not transform, attain the root of Dao." This is yet another example where the *Huangdi sijing* equates oneness with the root of Dao.

However, the fundamental issue is that the original Daoist vision of Dao, as much as that of heaven and earth, is not on a conceptual level. Although the *Laozi* states, "Dao is great, heaven is great, earth is great, humanity is great" (ch. 25), we cannot categorically say that the text places all these on the same conceptual level. The *Laozi*, besides emphasizing Dao, also stresses the importance of heaven, earth, and humanity as forces within Dao that form part of the myriad beings. Similarly, within the layered concept of the Three Forces, humanity does not occupy the same level as heaven and earth. Still, when the *Taiyi shengshui* insists that "the Dao of heaven values weakness," its notion of the Dao of heaven, as Li Ling and Ding Sixin put it, can be understood as analogous to the concept of Dao (Li 2002, 40; Ding 2002, 204). Among the conceptual levels of Daoist philosophy, Dao, oneness, and Great Oneness all occupy the top position, whether indicating the highest reality of cosmogony, the most fundamental meaning of life, the root cause of the myriad beings, or the essential underlying unity of all.

The *Daoyuan* further illustrates the relationship between Dao and Great Oneness as represented in the *Taiyi shengshui*. As its title "Source of Dao" indicates, the text explores the core of Dao. However, it contains many sections that do not speak of Dao at all. Rather, it seems to explore its opening line, "At the initiation of constant nonbeing, All was merged in great emptiness." Up to the statement, "Oneness is its appellation; emptiness is its dwelling; nonaction is its element; harmony is its function," Dao does not appear as the highest concept. However, this does not mean that the word "its" that modifies the four nouns does not in fact refer to Dao. There is no question that the opening line speaks of Dao, defining it in terms of Constancy Before and Great Emptiness, so it is quite likely that the latter statement also serves to characterize Dao. Later, the text says,

> Therefore, Dao is
> So high it cannot be scrutinized;
> So deep it cannot be fathomed.
> Clear and bright, no one can name it;
> Vast and spacious, no one can give it a form.
> It stands alone and is not paired with anything else.
> None among the myriad beings can command it. (see Yates 1997, 175)

This passage confirms that the word "its" after "oneness" in the earlier statement refers to Dao. Still, even without these lines and only using the title and its last section, which has, "Now, act in oneness and do not change; gain the root of Dao," it is obvious that in the *Daoyuan* oneness is Dao and Dao is oneness.

Dao is also a name for Great Oneness. When the *Taiyi shengshui* later says, "gray and murky are their names," the word "their" refers to Great Oneness as it manifests in heaven and earth. Some scholars have read the expression "gray and murky" (*qinghun* 青昏) to mean "may we ask" (*qingwen* 请问) (Li 2002, 39), but is more likely to mean "clear and murky" (*qinghun* 清昏). Seen from the way Daoists typically describe the original state of Dao or Great Oneness, "clear and murky" may well describe the as yet undifferentiated state of cosmic melee, where purity and turbidity are merged. The word "clear," moreover, can be related to the notion of Great Emptiness in the *Daoyuan*, while "murky" echoes the characterization of Dao as "vague and obscure" in the *Laozi* (ch. 21). In addition, the passage is comparable to the beginning of the *Hengxian*:

> In Constancy Before, there is no material existence. There is simplicity, stillness, and emptiness. Simplicity is Great Simplicity; stillness is Great Stillness; emptiness is Great Emptiness.

Furthermore, the *Taiyi shengshui* later says, "When the sage does things, he also depends on its name. Therefore, his achievements are made and his body suffers no harm." This also contains the phrase, "its name," most likely referring to the same subject. The text here means that people in general, and the sage in particular, should rely on the "clear and murky name" of Great Oneness. They should follow its example and go along with its original state. This is similar to the *Laozi*, "Common folk are bright and outgoing; I alone am murky and obscure. Common folk are discerning and discriminating; I alone am dull and confused" (ch. 20). It also echoes the *Zhuangzi*, "The essence of perfect Dao is dark and obscure; the ultimate of perfect Dao is murky and silent" (ch. 11). Darkness, obscurity, murkiness, and other, similar terms all emphasize that anyone pursuing Dao must enter a state of purity, clarity, and stillness. People as much as the sage must rely on the "clear and murky" as a way of naming this ultimate state and thereby getting closer to it.

Heaven and Earth

As regards the cosmogonic power and absolute nature of Dao or Great Oneness, the entire created world, including all forms and the myriad beings, is their counterpart. However, Daoist thinkers often single out heaven and earth for particular emphasis, indicating they have a special significance among the myriad beings. This is evident in the *Laozi*, as the following examples show.

> Heaven and earth are not humane.
> They treat the myriad beings like straw dogs. . . .
> Does not the space between heaven and earth
> Compare to a bellows?
> Empty yet never exhausted;
> Moving, yet always giving birth to more. (ch. 5)
>
> Heaven is eternal and earth everlasting.
> The reason why heaven and earth can be eternal and everlasting
> Is that they do not life for themselves.
> Therefore, they can be eternal and everlasting. (ch. 7)
>
> A violent wind does not last for a whole morning;
> A sudden rain does not last for the whole day.
> Who makes it so? heaven and earth.
> If even heaven and earth cannot make them last
> How much less can human beings? (ch. 23)
>
> Heaven and earth join together
> And send down sweet dew.
> Without human directions,
> It naturally reaches everywhere. (ch. 32)

Especially the first example, emphasizing that heaven and earth are not benevolent contrasts them with the myriad beings and places them into a position of the absolute. Still, in the overall philosophical structure of the *Laozi*, heaven and earth are still concepts of the first order. Dao or oneness are prior to them and give them birth—"There is a something, in chaos yet complete, ever since before heaven and earth were born" (ch. 25). They pervade heaven and earth without stopping—heaven attained oneness and became clear; earth attained oneness and became stable" (ch. 39). The same understanding also applies to the *Taiyi shengshui*. Here, too, Dao or Great Oneness are on a different conceptual level as heaven and earth.[8] They must not be confused.

---

[8] The *Zhuangzi* says, ""If we calculate the number of beings that exist, the count certainly does not stop at ten thousand. Yet we set a limit and speak of the myriad beings. This is because we select a number that is large, and agree to apply it to them. In the same way, heaven and earth are forms that are large, yin and yang are forms of vital energy that are large, and Dao is the generality that embraces them" (ch. 25).

From the perspective of the unfolding of the universe and all things and beings, heaven emerges from water, generated by Great Oneness; earth emerges from heaven, generated by Great Oneness. The generation of heaven and earth has a sequence in time, but after they have been created, they stand in a relative relationship to each other, yet are each in a different space (above and below) and of a different essence (soil and energy). Pervading the myriad beings, Great Oneness submerges itself in water—not unlike Laozi's statement, "All-pervading is Great Dao! It may be found left hand and right" (ch. 34). It continues to flow for ever, moving and revolving along with the myriad beings.

On the one hand, in order to maintain order in the universe, Dao and Great Oneness are counter posed to heaven and earth while the myriad beings still contain their fundamental nature. As the *Taiyi shengshui* says,

> It takes itself as the guideline of the myriad beings.
> This is
> What heaven cannot kill,
> What earth cannot bury,
> What yin and yang cannot form.

On the other hand, Dao and Great Oneness live by the principle of softness and weakness. This is why the text notes that "heaven values weakness." Laozi's philosophy places great emphasis on honoring the soft and weak. He repeatedly notes that they constitute the Dao's mode of action, recognizing that states of being strong and hard as well as full and overflowing can be harmful to the transformation and development of things. He believes that only by being soft and weak can one establish a balanced and harmonious relationship among things. In the *Taiyi shengshui*, the idea that the Dao of heaven values weakness finds expression among concrete things through heaven and earth achieving a delicate balance. Thus, it emphasizes that heaven values weakness and reduces completion, that it relies on weakness and strives to balance any excess or insufficiency.[9]

According to the above, core concepts of Daoism stand on different levels and in a hierarchical relationship. This hierarchy, moreover, provides a

---

[9] The *Liezi*, too, speaks of heaven slanting toward one side and Earth being insufficient: "When Gonggong was fighting Zhuanxu for the empire, he knocked against Mount Buzhou in his rage, breaking one of the pillars of heaven and snapping one of the threads that support earth. Therefore, heaven is excessive in the northwest, and the sun, moon, and stars move in that direction; earth is insufficient in the southeast, and the hundred rivers and rain floods flow home that way" (ch. 5; Graham 1960, 96).

particular expression of a unique logic. Certainly, the concepts of Dao and Great Oneness rank at the highest level of the entire system, despite the fact that they evolved differently. If we place notions of heaven, earth, and the like on the same level, this not only undermines the hierarchical relationship established in the *Taiyi shengshui* but also confuses the overall Daoist system of different existential layers in the universe. Philosophers inevitably will engage in debates, their thinking undergoing change and transformation. Still, the problem is that seeing Dao as a name for heaven and earth and not for Great Oneness does not match the fundamental outlook of the *Taiyi shengshui* and creates confusion and contradiction, constituting a serious error in our understanding of Daoist cosmogony.

School Affiliation

The last issue to address here is the question of where the text belongs in terms of early Chinese philosophical schools. To me, the *Taiyi shengshui* is a Daoist text, and most scholars agree with this assessment. Generally, philosophical schools tend to have clear affiliations even if they manifest in complex lineages. They all support certain fundamental beliefs and iconic concepts, and establish hierarchical relationships between certain key notions. Daoist philosophy in the broadest sense clearly follows this pattern, and even the rather narrow vision of the *Taiyi shengshui* matches it.

It contains key concepts of Dao and Great Oneness, central notions of water, heaven, earth, and the myriad beings, as well as a particular understanding of cosmic unfolding and a strong vision of heaven valuing weakness. All these form important building blocks of its ideological structure with its multiple levels and intricate divisions; they give its main doctrine and ideology a unique shape. What is more, they clearly reflect key characteristics of Laozi and early Daoist thought, being highly similar and compatible.

Conversely, materials pertaining to the thought of yin-yang cosmology do not contain Dao or Great Oneness as fundamental concepts, nor do they propose a systematic cosmogony or theory of cosmic unfolding. They do not emphasize Dao or Great Oneness as pervading the myriad beings, nor do they speak of the Dao of heaven as "valuing weakness." Examining yin-yang thought as represented in the *Shiji* and *Hanshu*,[10] and then again looking at

---

[10] They characterize the philosophy of the yin-yang cosmologists as follows: "Always focused on the arts of yin and yang, they emphasize ideas of good fortune and all sorts of taboos, causing people to live in fear of all kinds of things. They also

the *Taiyi shengshui*, it becomes obvious why the two cannot be compared. Therefore, the notion that the *Taiyi shengshui* should be classified as a document of yin-yang cosmology is untenable. Its thought belongs fully in the early stages of the evolution of Daoism, but even after realizing this, it is very difficult to draw a conclusion about its original author. Still, the most likely candidate is Guanyinzi.

In addition, I do not think that the *Taiyi shengshui* stands under a particularly strong influence from the school of numerology, as suggested by Peng Hao (2000, 540). This is because heaven and earth, yin and yang, the four seasons, spirit and light, hot and cold form fundamental concepts in all of early Chinese philosophy.[11] It is only natural that Daoist thinkers should use these concepts in describing the generation and order of the universe, but they read them in their own particular way matching their own particular cosmogonic vision. The *Huangdi sijing* provides two examples:

> Crowding together . . . make a single granary.
> Without darkness, without light,
> Yin and yang are not yet existeng.
> As yin and yang are not yet fixed,
> I have no way to name the myriad beings.
>
> Now, I begin by distinguishing them
> Into two and sepearat them into yin and yang.
> I divide them to make the four seasons. . . .
> This constitutes constancy. (*Shida jing*, ch. 2; Yates 1997, 107)

---

emphasize that the four seasons move in a strict order, which must not be lost" (*Shiji*, Pref.).

"The lineage of yin-yang cosmologists goes back to officials under Fuxi. They venerate the orderly movements of Highest Heaven, analyze the patterns of the sun, moon, planets, and stars, and have great respect for the people's seaons—all these constitute their key focus. Anyone following their teaching in practice needs to observe prohibitions and taboos, concern himself with specific numbers, work within human affairs, and go along with the demands of ghosts and spirits" (*Hanshu*, "Yiwenzhi").

[11] The *Yueque shu* records the story of Fan Li 范蠡, noting, "Dao emerged before heaven and earth. It never grows old but continues to bring forth the myriad beings. It cannot be appropriately named: thus, we call it Dao. Dao brings forth energy; energy gives rise to yang; yang brings forth yin; yin brings forth heaven and earth. Once heaven and earth are established, there are heat and cold, dry and wet, sun and moon, stars and planets, the four seasons, and all the myriad beings" ("Shenzhong" 枕中).

Heaven and earth, yin and yang,
The four seasons, the sun and the moon,
The planets and constellations, primordial energy,
The wrigglers that walk and crawlers that move,
Even the plants that grow roots—
They all take life from Dao,
But never decrease it.
They all return to Dao,
But never increase it. (*Daoyuan*; Yates 1997, 175)[12]

These passages contain several concepts essential to naturalist thought, such as light and dark, yin and yang, the four seasons, heaven and earth, planets and stars, clouds and vapor, etc. Most generally, all pre-Qin philosophers mutually criticized each other, but they also exerted a considerable influence upon each other, especially those who occupied a strong and proactive position, such as the Huang-Lao school. It integrated other schools of thought and also had a close relationship to numerology. Thus, naturalist thought pervades ancient Daoist cosmology, a similar strand of key concepts pervading all, from the *Taiyi shengshui* to the *Huangdi sijing*.

[12] The *Wenzi* contains two similar examples. "When heaven and earth were not yet formed, all was dark and obscure, merged into oneness. In silence, it began to become clear and pure. The heavy and muddy formed Earth, the essential and subtle formed Heaven. They separated to create the four seasons; they divided to make yin and yang. Their essential energy formed humanity; their coarse energy formed the animal kingdom. Hard and soft produced each other and the myriad beings came to life" (ch. 3).

"Dao brought forth the myriad beings; it structures itself through yin and yang. They transform and make the four seasons; they divide and establish the five phases. Each thus gets its particular place. They all come and go with the seasons, remain in constancy through rules and measures" (ch. 8).

Chapter Three

Fanwu liuxing

From Oneness to Multiplicity

Among the various excavated materials, the *Fanwu liuxing*, now at the Shanghai Museum, is relatively late, postdating the *Taiyi shengshui*, the *Hengxian*, and the *Daoyuan* of the *Huangdi sijing*. It presents a strong philosophical position and unique metaphysical outlook, centering largely on the concept of oneness. The word "Dao" occurs only three times in the text, including in the compound "Dao of heaven," but the word "one" or "oneness" appears on nineteen occasions, including twice the expression "one phrase." It, too, speaks of "holding on to Dao" or "oneness" and, very much like other early Daoist works, uses the two terms almost interchangeably. However, here Dao is clearly in a secondary position, superseded by oneness. For this reason, to understand its philosophical implications fully, it is essential to focus on this concept.

The Huang-Lao Connection

The study of ancient thought often focuses on certain core terms and concepts that form landmarks or watersheds in distinguishing particular schools and lineages. In this context, the notion of oneness is critical in differentiating different strands of Daoism, and especially its concrete visions within the Huang-Lao school.[1] For this reason, in studying the notion of oneness in the *Fanwu liuxing*, we must take its role in Daoism and Huang-Lao into consideration (see Wu 1985; Bai 1998).

---

[1] Other essential Daoist terms include Dao and virtue, however, they also commonly appear in Confucian works and do not have the same level of distinction. Oneness, on the other hand, both in form and substance, is a centrally Daoist concept—which, of course, does not mean that it never appears elsewhere. Within ancient Daoism, moreover, the philosophical elaboration of oneness is particularly the product of the Huang-Lao school, which is why we look at it especially within this context.

Huang-Lao thinkers are unique in that they take common numerals that also appear elsewhere as, for example, in *Zhuangzi* 33, and turn them into exceptionally abstract and highly metaphysical concepts. As for the number one, the *Shuowen* begins its work with it and explains it cosmologically: "In Pure Initiation and Great Beginning, Dao rested in oneness. Then it opened up, divided into heaven and earth, and gave birth to the myriad beings." This obviously goes far beyond the simple numerical meaning of "one," which is at the root of the concept and in its literal form occurs in the *Yinwenzi*: "All the numbers—ten, hundred, thousand, million, million, thousand, hundred, ten—arise from one" (*Taiping yulan* 750). One is the beginning of all numbers, but already the *Shuowen* describes it in metaphysical terms, reflecting a complex philosophical development and expansion. All transmitted Daoist texts play a role in this process, as do the newly excavated materials, supplementing and enhancing our understanding of its vision in the *Fanwu liuxing*.

Philosophically, Huang-Lao thinkers use oneness as another term for Dao. To them, Dao signifies the underlying unity and cosmic interconnectedness of universe, nature, and humanity, forming the central pillar that holds up its metaphysical edifice. The world that humanity faces is one of unlimited multiplicity: innumerable physical objects, variegated individual entities, and various phenomena in the natural world as well as never ending affairs and activities in society. From ancient times, human beings, gifted with a rational mind, have pursued the source of this amazing multiplicity, trying to understand the root cause of its multifaceted nature and hoping to come to grips with its underlying unity. Thus, Plato in ancient Greece speaks of forms and ideas in defining the common essence of things and describing the one behind the many. Plotinus specifically uses the term "Great Oneness" in discussing the first cause underlying the myriad beings (1989; 2004).[2]

In pre-Qin thought, Huang-Lao describes the root cause of the myriad beings and their underlying unity in terms of Dao as well as oneness and Great Oneness. Newly excavated materials show that Confucians were not particularly concerned with this question and did not offer any relevant answers. Huang-Lao thinkers, on the other hand, dealt with it in great detail, using oneness or Great Oneness to explain the inherent quality of underlying unity in the myriad beings and working with the numeral to show the singularity of this quality and thereby establishing a numerical connection between multiplicity as it occurs in the natural and human worlds. Since one,

---

[2] Naturally, his rather concrete idea of Great Oneness, as well as Plato's notion of oneness, are not at all the same as those proposed in Daoism. Plotinus's vision has a strong religious coloring, while Plato focuses heavily on the contrast to the many.

moreover, is the origin of all numbers, it also implies the notion of the beginning of all beings.

The school further describes the multiplicity of beings in the world within the framework of the Three Forces—heaven, earth, and humanity. Among them, heaven covers all, earth supports all, and humanity participates in all. In Huang-Lao thought, oneness appears as the highest, singular creative power, the original universal quality of life. It stands in contrast to the various concrete objects of the world, which are multiple and manifold. In other words, in contrast to the multiplicity of the myriad things in the universe, oneness represents their emergent and unifying force; in contrast to the manifold activities of the people in society, oneness as activated through the sage—the singular person—represents key aspects of political order and social unity. The core focus of Huang-Lao, we can therefore say, is the question of the relationship of the one and the many. This plays out on different levels and in various dimensions, as is shown in the following table of common terms used:

Universe

One	Many
Great Oneness, Dao, Constancy Before, Great Beginning, nonbeing, no form, root, cause, source, ancestor, mother, energy, simplicity, changes	material existence, forms, images myriad beings, differences, names, time, place, old, new, inherent nature, feelings, complexity, variety

Society

One	Many
great, noble, high, sage, king, ruler, lord, singular, alone, I, cause, flow, going along, stillness, clarity, emptiness, subtlety holding on to, embracing, recovering, guarding [oneness], one phrase, single mind, nonaction	small, humble, low, world, people, affairs, host, multitude, others, effect, moving, activity, muddiness, fullness, self-so, self-transformation, self-balancing, self-correcting, self-awareness, self-purity, self-being

The *Fanwu liuxing* uses these and other terms, as scholars have pointed out variously.[3] Examining its discussion of oneness in relation to the history of Huang-Lao thought, one can look at the big picture first and then examine the details, or vice versa. Either way, three major dimensions of oneness emerge: as the creative force of the universe, as the foundation of the exist-

---

[3] For editions of the text, see Fudan 2008; Gu 2009; Li 2008; Ma 2008; Taninaka 2010; and Wang 2009. Studies appear in Asano 2009, Cao 2010, and Wang 2010b. The English rendition used here is Chan 2014.

ence and activities of all beings, and as the internally structuring potential of the world. Let us look at these in turn.

Cosmogony

Oneness in the *Fanwu liuxing* serves first of all as the root source of the universe and the creator of the myriad beings. How does it function in this role? Called "oneness," it forms part of the overall Huang-Lao model of cosmogony, which in many ways echoes that of other Daoist texts, both transmitted and excavated, but adds its own unique twist to it. It says,

> I have heard,
> One gave birth to two.
> Two gave birth to three.
> Three gave birth to the mother.
> The mother gave birth to integration.

The first step in this system is fairly standard, naming "one" as the starting point and initial beginning and thus establishing it as the root source of the universe and creator of the myriad beings. However, it is not the same as the classic passage in *Laozi* 42, which begins with, "Dao gave birth to one" and has "three gave birth to the myriad beings." Here Dao is the root cause of all rather than one, which plays different roles in the *Laozi*.

> Keep the spirit and embrace oneness. (ch. 10)

> Thus, the sage embraces oneness
> And becomes the model for the world. (ch. 22)

> Look at it and do not see it: call it invisible.
> Listen to it and do not hear it: call it faint.
> Touch it and do not feel it: call it subtle.
> These three cannot be better understood,
> They merge and become one. (ch. 14)

> Those of old who attained oneness:
> Heaven attained oneness and became clear.
> Earth attained oneness and became stable.
> Spirit attained oneness and became numinous.
> Valley attained oneness and became full.
> Kings and barons attained oneness and became rulers. (ch. 39)

Both traditional commentators and modern scholars read "oneness" in most of these passages as Dao—"a being in chaos yet one" (ch. 25), a force that underlies the world but can be attained, embraced, and used. In chapter 10, "one" represents the union, the potent integration of the two types of souls (Chen 2003, 221, 161, 127, 109).

Xu Kangsheng describes oneness in Laozi as having three distinct meanings: Dao, the unity of things created by Dao, and the body (1985, 9; see also Jiang 198, 78). I see it as being two things, a substitute term for Dao and the initial state of unity at the dawn of creation. In this sense, the term also appears in the *Zhuangzi*, "In Great Initiation, there was nonbeing; there were neither material existence nor names. Out of it oneness arose. There was oneness, but it had no form. Beings attained it and took birth: it was called virtue" (ch. 12). In this model, the beginning of all was before material existence and names, while oneness appeared as a second stage, after Dao or Great Initiation. It is not just an alternative way of referring to them.

Unlike this, oneness in the *Fanwu liuxing* is also Dao, and Dao is also oneness. The two terms are used interchangeably, as is generally the case in Huang-Lao works, both signifying a high metaphysical concept. There are, however, certain passages where oneness serves to characterize certain dimensions of Dao. For example, the *Shida jing* of the *Huangdi sijing* has, "Oneness is the root of Dao" (ch. 9); the *Hanfeizi* uses the term to emphasize that Dao is "one and not two." It notes,

> Dao differs according to the myriad beings; virtue differs according to yin and yang; weight differs according to light and heavy; rope differs according to push and pull; climate differs according to dry and damp; rulers differ according to masses and ministers. All these six emerge from Dao. Dao is not two: thus we speak of it as oneness. (ch. 8)

These passages are the exception, however. For the most part Huang-Lao works use Dao and oneness in the same manner or describe oneness as the original state of Dao. For example, the *Daoyuan* says,

> Oneness is its appellation;
> Emptiness is its dwelling.
> Nonaction is its element;
> Harmony is its function. (see Yates 1997, 175)

The *Wenzi* states, "Dao has no form and no sound, therefore the sage gives it form as strength and a name by calling it oneness" (ch. 2, "Jingcheng" 精誠). In these two statements, oneness serves as the name or appellation of Dao.

This is also very much like the *Taiyi shengshui*, where Great Oneness stands for Dao, and the *Daoyuan*, which says,

At the initiation of constant nonbeing,
All was merged in great emptiness.
Empty and same, all was in oneness,

The *Lüshi chunqiu* echoes this, "Dao is perfect essence: beyond forms and names. Forced to give it an appellation, we call it one" (5.2). The *Huainanzi* notes, "Cavernous and undifferentiated heaven and earth, chaotic and inchoate uncarved block, not yet created and fashioned into beings: this we call Great Oneness" (14.1; Major et al. 2010, 536). The modifier "great" emphasizes its highest, ultimate nature.

The *Fanwu liuxing* expresses the same idea. Since oneness is an easy way to express the unified nature before division of the prime cosmic state and since it is a straightforward term to use in contrast to the multiplicity of the myriad beings, the text not only uses it as an alternate for Dao but also to represent its basic nature. It uses "oneness" more than "Dao" to set up its particularly metaphysics and through it defines its cosmogonic starting point. The same idea is also present in the *Wenzi*, "Merged, it becomes one" (ch. 3); in the *Lüshi chunqiu*, "Great Oneness gave rise to the two forces" (5.2); in the *Taiyi shengshui*, "Great Oneness;" in the *Daoyuan*, "Empty and same, all was in oneness;" and in other passages of the period.

One universal aspect of being human is curiosity about our origins, an awareness of our source. Present from the dawn of history, it has led to various visions of our "original" or "natural" state as well as to imaginary scenarios of what the universe was like without us. In Huang-Lao, we can see two manifestations of this, expressed particularly in the *Fanwu liuxing* and its description of what the ultimate state of oneness was like. However, it also occurs in various other Daoist materials that present alternative visions.

Laozi presents Dao as the primordial force of the cosmos.

Dao is pervasive.
Use it, but its capacity is never exhausted.
Bottomless, it seems the ancestor of the myriad beings.
Deep and still, it appears to exist forever.
I do not know whose son it is.
It seems to have come before the gods. (ch. 4; Chan 1963, 141)

Dao as being is elusive and vague,
Vague and elusive: in its midst is an image.
Vague and elusive: in its midst are beings.

Dark and obscure: in its midst is essence.
Essential and perfect: in its midst is trust. (ch. 21)

Along the same lines, the *Hengxian* speaks of the original state as follows,

> In Constancy Before, there is no material existence. There is simplicity, stillness, and emptiness. Simplicity is Great Simplicity; stillness is Great Stillness; emptiness is Great Emptiness.

The cosmogonic model of the *Fanwu liuxing* describes oneness as the starting point and self-being or naturalness as the original state, but the text does not provide a detailed description.

Cosmic Unfolding

The process of cosmic unfolding in Huang-Lao thought tends to work from chaos to division, from the simple to the complex, from the few to the many—not unlike the universal unfolding described by Spencer (see Bowler 1989; 1999, 312). The *Fanwu liuxing* is unique in its strong emphasis on oneness as the initial condition and starting point of emergence, but it still follows the overall model of moving from the one to the many.

It is also special in its choice of terms. While the movement is from one to two to three, it is close but not identical to Laozi's classic passage. It is close in that it uses the same rhythm of three-character phrases (X gave birth to Y) and in that it seems to use the same basic stages, unlike any other model presented in early Chinese thought. It is different in that it posits different beginning and end points—oneness rather than Dao, mother and integration rather than the myriad beings. In addition, it also uses different variants for "two" and "three" using *liang* 两 for *er* 二 and *can* 叁 for *san* 三. The meaning remains the same, though, and it is an open question whether the variants carry a special significance.

The tradition commonly associates the stages of the two and the three in Laozi's cosmogony with energy: yin and yang plus their joining in harmony. The latter gives birth to the myriad beings. A first clear formulation of this appears in the *Huainanzi*:

> Dao begins with oneness. Oneness alone, however, does not give birth. Therefore, it divided into yin and yang. The harmonious union of yin and yang gave birth to the myriad beings. Thus, it is said, "One gave birth to two. Two gave

> birth to three. Three gave birth to the myriad beings." (3.28; see Major et al. 2010, 133)

"Two" here signifies the energies of yin and yang, while "three" indicates the two energies in harmony.[4]

The *Liezi* describes cosmic unfolding as moving from Great Change through Great Initiation and Great Beginning to Great Plainness: "Great Change preceded the appearance of energy; Great Initiation was the beginning of energy. Great Beginning was the beginning of form; Great Plainness was the beginning of substance" (ch. 1). It thus works with four distinct stages that characterize the movement of oneness. As it continues,

> Oneness is the beginning of all form and transformation. The clear and light rose up to become heaven; the muddy and heavy sank down to become earth; the blended and harmonious energy became humanity. For this reason, heaven and earth contain their essence, and the myriad beings transform and take birth.

Unlike these works, the *Fanwu liuxing* uses the term "energy" (*qi*) only once, when it says, "The five energies began to arrive." What, then, are the "five energies"? The *Zuozhuan* speaks of the "six energies" of yin, yang, wind, rain, darkness, and light (Zhao 1). Its commentary notes, "The five phases are the five energies." Cao Jinyan 曹锦炎 thinks the *Fanwu liuxing* does indeed speak of the five phases. If this is the case, these five are still not the same as the yin and yang energies discussed in the *Laozi*.

The *Fanwu liuxing* also contains the notion of yin and yang, asking, "The order of yin and yang—how did it get established?" In addition to discussing things in terms of "energy," it speaks of yin and yang but does not connect them explicitly with the arising of the myriad beings. All this casts some doubt on whether "two" and "three" indicated in the text are the same as those mentioned in the *Laozi* and whether the text matches the common vision found in Daoist mainstream thought.

It continues its outline of cosmic unfolding by stating that "three" gave birth to the "mother," which in turn gave rise to "integration." The mother, as is well known, is a common Daoist metaphor of cosmic origin, applying the human relationship of mother and children to universe and myriad beings. However, rather than being at the very source of existence, the mother in the *Fanwu liuxing* only appears on stage three, quite different from Laozi's use of the metaphor, which applies to Dao at the very beginning of all. It is

---

[4] The same explanation also appears in the *Wenzi* (ch. 3), but without the concrete contrast.

also unlike the *Taiyi shengshui*, which speaks of Great Oneness as the "mother of the myriad beings." Here, mother signifies the intermediate stage right before the emergence of living beings—quite possibly matching what other texts describe in terms of heaven and earth.

Commonly in Daoist texts, heaven and earth appear as a major stage where there is "form" for the first time, i.e., the point from which the myriad beings first emerge. They are accordingly called "the mother of the myriad beings," an example being *Zhuangzi* 19. The *Fanwu liuxing* makes no direct reference to heaven and earth in this role, yet it emphasizes their unique and special position. It further places mother at the next-to-last level, the final stage being "integration," which sees the emergence of the myriad beings.

The myriad beings are the final stage of cosmic unfolding not only in the *Laozi*—"Three gave birth to the myriad beings"—but also in other early texts. Thus, the *Lüshi chunqiu* has,

> The myriad beings were born from Great Oneness. . . . Great Oneness gave birth to the two forces; the two forces gave birth to yin and yang. Yin and yang transformed, one going up, the other going down. They combined and became complete . . . The myriad beings emerged from them. They were born from Great Oneness and transformed through yin and yang. (5.2)

They also form the final outcome of creation in the cosmogonic model of the *Huainanzi*, but here the starting point is not oneness but Dao.

> Dao began in emptiness and vastness. Emptiness and vastness gave birth to space-time. Space-time gave birth to original energy. A boundary divided original energy. That which was clear and bright rose up and spread out to form heaven; that which was heavy and muddy sank and congealed to form earth. It was easy for the clear and subtle to converge but difficult for the heavy and muddy to congeal. Therefore, heaven was completed first; earth was fixed afterward. The conjoined essences of heaven and earth produced yin and yang. The supersessive essences of yin and yang caused the four seasons. The scattered essences of the four seasons brought forth the myriad beings. (3.1; Major et al. 2010, 114-15)

The *Fanwu liuxing* is exceptional in that it does not place the myriad beings at the culmination of cosmogony, but rather speaks of "integration." The word for "integration" (*jie* 结) originally means "knot" or "bond." Qiu Hualin (2009) cites the *Heguanzi*, "So, divine luminaries firmly tie their bonds. Every race born and complete will employ oneness inexhaustibly" (ch. 11, "Tailu" 泰錄; Wells 2013, 168). This, too, uses the term in the sense of "assembly" or "combination." If the word essentially means "assembly" or "integration." its opposite is "dispersal" or "multiplicity," the result of the creative process. The fourth stage in the cosmogonic model of the *Fanwu liuxing* being "integration," then, the emergence of the myriad beings is the level when the "various beings are flowing in form." "Integration" means "completion," reaching a stage of togetherness. As the *Zuozhuan* notes, "Using yin to integrate it" (Xiang 12). The integration, the completed presence of "the mother of heaven and earth," can thus be understood as the level when the myriad beings emerge.

The Natural World

The second major dimension of oneness in the *Fanwu liuxing* is that it serves as the foundation of the existence and activities of the myriad beings. As noted earlier, Huang-Lao thinkers describe Dao, oneness, Great Oneness, and Constancy Before as the root source of the emergence of the myriad beings, characterizing them as "manifold" and essentially meaning the various physical objects and living beings of the natural world.

Oneness is the creator, while the many are the created. In Huang-Lao Daoism, this contrast is also expressed in terms of formless versus formed, imageless versus imaged, nameless versus named, never impacted versus potentially impacted, and the like. Both the *Liezi* (ch. 1) and the *Zhuangzi* (ch. 11) use related terms, contrasting "that which gives birth" and the "born," "that which gives form" and the "formed," as well as "that which makes beings beings" and "beings." Entities that have form, image, name, and the potential to be impacted are called beings or the myriad beings. They all exist in individual bodies and manifest as phenomena in the world. The *Fanwu liuxing* describes them as many and characterizes them as things that have shape and color and come in great variety.

The first half of the text consists largely of questions about the formed entities and natural phenomena that make up the world. It is reminiscent of Qu Yuan's *Tianwen*, however, unlike the latter, the text here provides answers in its second half, switching to a narrative, explanatory mode, introduced with the phrase, "I have heard it said." The very structure of the text, thus, echoes the universal pattern of the one versus the many, creation versus created, origin versus existence. How, then, does the text speak of the various entities that populate the world?

To begin, it asks about their origins, inquiring into the "why" of their existence, in a long string of over forty questions for which it does not provide direct answers. The questions reflect its author's observations, concerns, and reflections regarding the natural world.[5] Many deal with the abstract notion of "beings" or the "hundred beings"—terms used as a comprehensive appellation for all the various things in the world. Ancient texts most commonly speak of the "myriad beings," but some also use expressions like "various beings" or "all beings."

Beings in general live and change by manifesting in "form" (*xing* 形) and "organism" (*ti* 体)—terms used differently in the text. "Form" here indicates the condition of beings when they are first born, while "organism" means the states they undergo as they grow and mature. The text begins, "All things flowing in form—how can they attain completion? Flowing in form and growing into organism—how can they not die?" "Flowing" is a term taken from the natural world and indicating the movement of water. In this context, it means the ongoing transformation and individual substance of beings, not unlike our term "develop." The word also occurs in the *Yijing*, "Various beings appear flowing in form" (Hexagram 1; Sung 1971, 3); and the *Shijing*, "Now they are rich; now they have organisms" ("Daya"). "Completing their organism," thus, means that beings reach maturity in their bodily substance.

Going beyond beings, forms, and organisms, moreover, the text questions the natural world in two aspects. First, it asks about heaven and earth; second, it questions the nature of the beings between heaven and earth. Heaven and earth, as described above, indicate the largest, most obvious and most ubiquitous of beings, sometimes also described as agents of emergence, containing an inherent quality of primordial origin. Here, the text notes, "When heaven and earth established the end and the beginning, heaven sent down five measures." This must refer to the beginning and end of the myriad beings plus ways of following Dao sent down from heaven. This shows that the text places heaven slightly ahead of earth and sees it as more original.

It then raises two questions, "Why is heaven high and earth far-reaching?" and, "Who made heaven; who made earth?" Recognizing that heaven and earth have qualities of origination, the text thus also sees them as created themselves, asking how they came to be and looking for their root source. Next, it questions the world defined through heaven and earth, detailing specific entities for each. Sun and moon, thunder and lightning, wind and rain belong to heaven; water and fire, grasses and trees, birds and beasts, soil and people, ghosts and spirits belong to earth. Aside from ghosts and spirits, these are all things human beings commonly recognize as part of the

---

[5] Similar questions also appear in the *Zhuangzi* (ch. 14) as well as in the *Zhouyi shu* 周逸书. See Cao 2009.

natural world—that change and develop and appear in different modes in the course of their lives.

Moving on, the text asks about the cause of natural beings, phenomena, and transformations:

> Why do grasses and trees grow? Why do birds and beasts cry? Why is the earth flat? Why do rivers flow clear? Why do people live and die? Why are there wind and rain, thunder and lightning? Why, when the sun first rises, is it big and not yet hot? Why, when the sun sets, does it turn to be small?

These are common natural phenomena, familiar to human beings, seen on a daily basis and usually accepted as self-explanatory. The questions the text asks about them in modern terms would fall in the range of physics, a field in antiquity not delimited strictly from philosophy.

Wwe often know what certain natural phenomena are but we do not know why they are what they are. The same holds true for people of old. The questions of the *Fanwu liuxing* therefore reflect an inherent human curiosity and insecurity vis-à-vis natural objects and phenomena.

The Foundation of Beings

Daoist explanations of the universe and the natural world tend to be sweeping and overarching, not only presenting models of how the myriad beings came to be but also why. For example, the *Huainanzi* says that the myriad beings emerged from Great Oneness and manifested as birds, fish, animals, and the like, providing an explanation of how the multifaceted world came about from a single origin. It says,

> Together emerging from oneness, so that each acquired its distinctive qualities, there were birds, there were fish, there were animals: this we call the 'differentiation of things.' Regions became distinguished according to their categories; beings became differentiated according to their groupings. Their inherent natures and destinies were dissimilar; yet all acquired their physical forms in the realm of material existence.
>
> Separate and not interconnected, differentiated as the myriad beings, none could return to their ancestor. Thus, when animated, beings are said to be alive; when dead, they are said to be expired. In both cases, they are beings. It is not that there was nothing that made them into beings; rather, what made them into beings is not among the myriad beings. (14.1; Major et al. 2010, 536-37)

This offers an overarching vision of the foundation of beings' existence and activities, a tendency common in Huang-Lao metaphysics, which usually speak of Dao and oneness in this context. The *Laozi*, too, uses oneness as the key factor, explaining how heaven, earth, spirit, valley, and kings got to be clear, stable, numinous, full, and ruling (ch. 39).

Huang-Lao texts frequently use the term "attain" or "get" (*de* 得), also applying it to their explanation of how various beings and phenomena became what they are.[6] For example, the *Daoyuan* says when discussing how Dao affects the multiplicity of the myriad beings,

> It takes oneness as its measure and never changes,
> And is right there even for crawling insects.
>
> Birds attain it and fly.
> Fish attain it and swim.
> Wild animals attain it and run.
> The myriad beings attain it and live.
> The hundred affairs attain it and are complete.
>
> All men use it,
> But no one knows its name.
> All men employ it,
> But no one sees its form (Yates 1997, 173-75)

The *Zhuangzi* notes, "Moreover, Dao is the path by which the myriad beings proceed. All beings that lose it, die; all that attain it, live. To go against it in one's undertakings is to fail; to comply with it is to succeed" (ch. 31). The *Hanfeizi* speaks of "getting" it, emphasizing the universal nature of Dao:

> Heaven can be high because of it; earth can hold everything because of it. The pole star can have its majesty because of it; the sun and the moon can make constant illumination because of it. The five phases can keep their positions constant because of it; all the stars can keep their right orbits because of it. The four seasons can control their diverse expressions because of it; Xuanyuan could rule over the four directions at his discretion because of it. Master Redpine could live as long as heaven and earth because of it; and sages can compose essays and elaborate institutions because of it. . . .
>
> Because of it, the myriad beings die; because of it, they live. Because of it, all kinds of affairs fail; because of it, they succeed. Dao can be compared to water.

---

[6] Wang Bi comments on *Laozi* 39, "One is the beginning of all numbers and the ultimate power of beings. In each case, it provides for the birth of the being and makes it perfect. All beings get the One to be complete."

Who is drowning dies as drinks too much of it. Who is thirsty lives on as he drinks a proper amount of it. Again, it can be compared to a sword or a speak. If the stupid man uses it for wreaking his grudge upon others, calamities will happen. If the sage uses it for punishing the outrageous, good fortune will ensue. Thus, people die of it, live owing to it, fail because of it, and succeed on account of it. (ch. 20; Liao 1959, 193)

In comparison, the *Fanwu liuxing* presents oneness as the foundation of beings in two ways. First, it emphasizes oneness as the ground of all existence—as long as there is oneness, the world exists in all its complexity; without, it would not be there. "Therefore, when there is oneness, there is nothing under heaven that cannot come into existence; if there is no oneness, there is nothing that can exist." Second, it is the root cause of all natural manifestations, the growth of plants, the cries of animals, and so forth: "Grasses and trees attain it and grow; birds and beasts attain it and cry."

These statements in the text serve to answer the questions it raises in its first half, "Why do grasses and trees grow? Why do birds and beasts cry?" Seen from this angle, the text uses oneness to explain all the different aspects of reality, including the cycle human of life and death. "How come, after their flowing form has grown into organisms, they lose it and die? They attain it and find completion without ever knowing the realities of left and right." According to the original editor, Cao Jinyan, the expression "left and right" here refers to relative positions in the world. Liao Mingchun reads it to indicate leadership and control (2008), while Cao Feng sees it as "two sides of various circumstances" (2009).

Other questions arise: What exactly is it that beings lose when they die? Why do they become complete as they attain it? Reading the phrase as indicating leadership or creative potential makes a lot of sense. The word for "realities" (*qing* 情), moreover, echoes its usage in the *Zhuangzi*, "You do not understand the principle of heaven and earth, the reality of the myriad beings" (ch. 17). The *Shiwen* 十问, unearthed at Mawangdui, similarly speaks of "examining the realities of heaven and earth."

Why is oneness or Dao sufficient as the root and foundation of the existence and activities of the myriad beings? Simply put, it is because of its transcendence and infinity. Huang-Lao works often uses negative expressions to describe this: nameless, imageless, soundless, flavorless, unmoved, untouched, etc. Classical passages include the following:

Look at it and do not see it: call it invisible.
Listen to it and do not hear it: call it inaudible.
Touch it and do not feel it: call it subtle. (*Laozi* 14; Chan 1963, 146)

As for Dao,
It is what the mouth cannot speak of,
The eyes cannot see,
And the ears cannot hear. (*Guanzi*, "Neiye;" Roth 1999, 56)

Formless: that is oneness.
Oneness is being merged in no-mind with heaven and earth.
Spread its virtue but don't rouse it,
Use it but make no push it,
Look at it but do not see it.
Listen for it but do not hear it. (*Wenzi*, ch. 1)

The *Fanwu liuxing* follows this, but with one major difference. Although it acknowledges the transcendent and absolute existence of oneness or Dao, it also sees it as something people and beings can directly perceive and experience.

Therefore,
Oneness can be tasted when chewed.
Its scent can be perceived when smelled.
It makes sound when clapped.
It can be seen when approached.
It can be handled when managed.

But
It is lost if grasped tightly.
It withers if suppressed.
It is extinguished if opposed. (Chan 2014)

This description of oneness is highly unusual and unique to the text, significantly different from other Huang-Lao materials. However, the *Zhuangzi* contans a passage that points in the same direction. Master Eastwall asks Zhuangzi, "What we call Dao, where it is?" Zhuangzi answers that it is present in ants and crickets, grass and weeds, tiles and shards, and—when pushed further—even in excrement (ch. 22; Watson 1968, 241). Eastwall thinks of Dao as a concrete thing and looks for it in different places, while Zhuangzi's Dao goes beyond that and indicates an underlying structural, inherent quality of all things. This inherent quality is present everywhere and cannot be pinpointed as one specific entity. The *Zhuangzi* continues,

That which treats things as things is not limited by things. Things have their limits - the so-called limits of things. The unlimited moves to the realm of limits; the limited moves to the unlimited realm.

For Zhuangzi, Dao is everywhere, in each and every concrete entity, no matter how small or insignificant. All existing beings have Dao as their inherent quality. This also means that it is quite impossible to perceive Dao directly or separately.

The *Fanwu liuxing*, too, sees Dao as the inherent quality and underlying creative power of all that exists, yet here it is possible to perceive or intuit its presence. Though mysterious and wondrous, Dao can be expressed metaphorically in forms and images; it can become extremely close and very familiar. As the *Hanfeizi* notes,

> Dao . . . when close to me yet makes me roam through the four extremes. When far off, it is yet my constant principle. When dark, it radiates with amazing brightness. When light, it is mysterious and obscure. It bring merit all over heaven and earth, gives birth to thunder and lightning. All beings in the greater universe hold on to it for their completion. (ch. 20)

All this goes to show the special worldview expressed in the *Fanwu liuxing*, quite unlike anything else in Huang-Lao works, which tend to emphasize the impossibility of perceiving or managing the underlying power of the universe.

The Sage King

The third dimension of oneness in the text has to do with the singularity of the sage king in contrast to the multiplicity of the people, one lord versus many subjects. Oneness here is the dominant principle the sagely ruler holds on to create order in the world: oneness as the foundation of the arts of rulership. Defining oneness as the root source and inherent quality of all existence, how does the text apply this to politics and government?

Huang-Lao thought in general sees social structure and political order in close connection to cosmic patterns, the microcosmic expression of a universal system. Political oneness, therefore, is an expansion of oneness as a natural principle; human law is an extension of the law of Dao—just like Dao in Laozi's metaphysics and his practice of nonaction are guidelines for social order and ways of governing. In Huang-Lao, all civil laws, statues, and regulations come directly from Dao, but its application in the human world is not the same as in nature. In the natural world, oneness gives birth to and completes the multiplicity of the myriad beings as an internal force. In society, it has to be mediated by a high-level personage—the sage as ruler over the people.

The *Fanwu liuxing* matches overall Huang-Lao thought in that it uses oneness as a principle of governing, but when it comes to its concrete application it goes its own way. For example, it does not draw a connection between oneness and concepts like laws and measures, emptiness and stillness, or nonaction. Leaving aside certain radical thinkers who proposed models reminiscent of republicanism and anarchism, ancient Chinese philosophers typically saw a divide between the ruler and the ruled and assigned specific roles to each—as, for example, Mencius's notion that the ruler labors with his heart-and-mind, while the ruled labor with their physical strength. Huang-Lao works for the most part with the sage as ruler or overlord and the people or masses as the ruled.

So far, the schema is entirely the same as in classical Confucianism. However, in Huang-Lao, the relationship between the sage ruler and the people is a reflection of the abstract connection of oneness to multiplicity applied to the political order.[7] The contention is that the singular figure of the ruler or sage relies on certain inherent abilities to effectively create order and govern the masses. The *Shida jing* describes it when it tells the story of the Yellow Emperor asking Li Mo what kind of "complete law" the "only one or single person" possesses "to correct the people" (ch. 9). The answer is, "oneness." It serves as this kind of "complete law." The same idea is also apparent in the *Wenzi*, where King Ping asks, "How many ways of ruling does the king have?" The answer is, "The king has one."

The general position of Huang-Lao is that the ruler represents oneness in human society. In order to govern the innumerable multitude of the people, he has to hold on firmly to the treasured law, which is oneness in action. Huang-Lao texts speak of this as "holding," "guarding," "attaining," or "embracing" oneness. In other words, they see the ruler in human society as joined or merged with cosmic oneness, a fact that is part of who the leader is. By the same token, they make it clear that the political leader who centers himself on Dao—the sage, sage king, or enlightened ruler—is the one to hold on to the key principle and not any ordinary person.

Oneness (or Dao) here is the central principle of governing, the core of the arts of rulership, accessible to the sage or ruler. Even more than that, the sage or ruler is the physical, personal embodiment of this principle. Here lies the main divergence to Confucianism. Confucians see the sage king as a mod-

---

[7] The sage as singular is expressed clearly in *Laozi* terms, such as "alone," "orphaned," or "starved" (ch. 39).

el of Dao and virtue; Huang-Lao thinkers see him as an archetypal figure who holds on to (and even embodies) oneness.

How, then, does the *Fanwu liuxing* see the concrete activation of this? It has a character that does not exist in modern writing, 𢦏. The word consists of the characters for "little" (*shao* 少) over "head" (*shou* 首), combined with "lance" (*ge* 戈) on the right. The original editors of the manuscript read this to mean "be aware" or "know" (*shi* 识). I personally like to read it "hold" (*zhi* 执), following Yang Zesheng (2009; see also Wang 2009). The Guodian manuscripts distinguish clearly between these characters; their "hold" is similar but not identical to the mystery word in the *Fanwu liuxing*. "Holding on to" Dao or to oneness is a common thread in all early Daoist materials, often used interchangeably with "attaining" or "guarding" it and acknowledged as the highest principle of governing. The *Fanwu liuxing* uses these terms as well as "attaining," "having," and "not having oneness," but it does not offer a clear description of who exactly is supposed to do so. [8]

According to the most documents, holding on to and applying Dao or oneness in society is not something ordinary people do, but reserved for the political leader, the sage or ruler. They typically make three points. First, they speak of "holding on to Dao" in conjunction with applying self-cultivation to ordering the state, linking the personal connection to Dao with the power of government. "Holding on to Dao" means to govern; that is, the person who does so is a social and political leader. Second, they speak of "holding on to Dao while seated and never leaving the mat" and insist that "the sage remains in one place," which implies working in administration rather than by physical labor. Third, they use the expression "what the mind and heart holds dear is only oneness" when they speak of the ruler. All this goes to show that a close connection to or even merging with Dao is the characteristic of the leader and not of ordinary people.

Social Application

How, then, is the sagely leader supposed to apply this in everyday reality? Huang-Lao texts typically associate "holding on to oneness" with certain key

---

[8] Yanaka Shinichi discusses this issue with particular reference to the *Fanwu liuxing* in a paper he originally presented at the 2010 Conference on Ancient Chinese Manuscripts, held at the Center of General Education at Chia Nan University of Pharmacy and Science. See Yanaka 2010. I am much indebted to his rich findings.

concepts, such as nonaction, no desires, going along, clarity, stillness, one phrase, unified names, single-mindedness, and focused intention, in combination with overarching and firm centralized laws, statutes, and regulations. Huang-Lao developed through the joining of Daoist oneness with Legalist law and sees oneness as concretely expressed in human laws and social regulations—a connection entirely lacking in the *Laozi*.

When the sage king in Huang-Lao, therefore, "holds on to oneness," he in fact holds on to the "one law." The *Shida jing* says accordingly, "I have heard about the complete laws of the world. Therefore it is said, 'Do not use many, just speak one phrase and then stop. Follow names and go back to oneness, and the people will not fail to obey the standards" (ch. 9; Yates 1997, 135). The *Heguanzi* notes, "That which guards oneness in Dao and thereby controls the myriad beings is the law" (ch. 8, "Duwan" 度萬; Wells 2013, 134); and, "Oneness becomes law to complete its enterprise, so that none do not follow Dao. When oneness as law is established, the myriad beings all come in allegiance" (ch. 5; 2013, 109).

In other words, Huang-Lao Daoists take the abstract notion of Dao or oneness and make it concrete, socially applicable as the one law. This means, the ideal of the sage holding on to oneness changes into the laying down, upholding, and administration of specific codified laws. Here the Huang-Lao transformation of Laozi is at its most radical. As the *Yinwenzi* says,

> Firmly establish names and divisions, and the myriad affairs will not be in disorder. For this reason, people establish measures to extend long and short, amounts to receive much or little, weights to even out light and heavy, laws to judge clear and muddy, names to determine fake and real, laws to settle order and disorder, rules to guide the troubled and misled, adjustments to navigate dangers and difficulties. Doing so, the myriad affairs all return to oneness, the hundred measures are all standardized in law. Return to oneness is the ultimate of rules; standardization in law is the ultimate of adjustments.

This makes it clear: Huang-Lao thinkers see firm, codified and standardized laws and regulations as the key to "holding on to oneness"—but in combination with all the other concepts listed earlier, such as nonaction, no desires, and so on. There is no conflict here between personal quietude in an easy approach to rulership and the strict observance and enforcement of laws: holding on to nonaction *is* holding on to laws and regulations. In fact, the ruler is able to practice nonaction precisely because he consistently maintains set laws and regulations for the people and is superbly effective in their application. That is to say, because standardized laws are universally applied, the ruler can relax in nonaction.

All this applies to Huang-Lao documents in general. The *Fanwu liuxing* does not speak of nonaction, no desires, or any of the other key concepts, nor does it discuss laws and regulations, but it does contain related ideas. Its main concrete notions are "one phrase" (*yiyan* 一言) and "few words" (*guayan* 寡言).The former is a central concept in Huang-Lao political thought. Confucius already used the phrase, "The *Shijing* contains three hundred poems, but they can be covered in one phrase" (*Lunyu* 2.2); and, "May there not be expected from this one phrase the prosperity of his country?" (13.15). Here "one phrase" indicates a few words of major importance. Similarly, when Laozi speaks of governing through nonaction, he emphasizes "practicing the teaching without words" (ch. 2), notes that "many words lead to swift exhaustion" (ch. 5), and insists that "self-being has subtle words" (ch. 23). He does not specifically use the expression "one phrase," but the concept is clearly there.

Other texts are more explicit. The *Shida jing*, as cited earlier says that one should speak only one phrase and stop. "Now, a hundred words have a basis, a thousand words have their generalities" (ch. 9; Yates 1997, 137). Similarly, the *Guanzi* notes,

> Hold fast to oneness and do not lose it,
> And you will be able to rule the myriad beings.
> Rulers act upon beings,
> They are not acted upon by them,
> Because they grasp the guiding principle of oneness.
>
> With a well-ordered mind within you,
> Well-ordered words issue forth from your mouth,
> And well-ordered affairs are imposed upon others.
> Then the world is well ordered.
> When one phrase is attained,
> The world will submit.
> When one phrase is set,
> The world will listen
> This is what this means. ("Neiye;" see Roth 1999, 62, 64)

In Huang-Lao texts, "one phrase" is also expressed as "few words," "subtle words," or "minor words." In the *Fanwu liuxing*, these expressions refer to the practical application of oneness in the world.

> Thus I have heard,
> One phrase is never exhausted even to the end.
> One phrase attracts the masses of the people.
> One phrase benefits the multitudes.

> One phrase represents the purpose of heaven and earth. . . .
> He who can use few words matches oneness—
> This is what we call basic achievement. (see Chan 2014)

The first part of this closely echoes the *Guanzi*, while the latter is more like *Shida jing* (ch. 15).

The sage as one holding on to oneness stands in contrast to the many, i.e., the people, the masses, or common folk. The *Fanwu liuxing* states, "One phrase benefits the multitudes," giving expression to the actions of the one person at the pinnacle of society, whose actions will be advantageous and helpful to the common people. It also says, "That which the people value the most is the single ruler," again making the contrast between one and many obvious. Then again, it states, "If you are able to hold on to oneness, the hundred beings will not be lost," connecting the multitude of beings to the people as many. Other uses of "many" in the text relate it to the masses, the world, and heaven and earth.

In all cases, the idea is one of large numbers, of multiplicity and variegation, indicating the multifaceted nature of beings and phenomena in the world. Related expressions include hundred beings, different forms, different abilities, seasons, old and new, realities, and transformations—all implying changing landscapes in a sea of multitudes. The people as many, moreover, are often related to issues of the world, hundred affairs, myriad words, and the like. Specific terms are few—heaven and earth, world, domain, state, four seasons, thousand miles—but they all have to do with ways and areas of governing.

Traditionally the ruler faces the innumerable activities of large numbers of people. In Huang-Lao thought, the only way he can govern is through oneness, because only by working with the underlying principle of the universe can he effectively bring order to all existence without exception. This means that the core principle of governing is "using oneness to order the many," which furthermore connects to the political concept of all-pervasive Dao. This is common in Huang-Lao, which is why Sima Qian 司马迁 says that Daoists "point to connections and easily grasp things, have few affairs and many merits" (*Shiji*, pref.), why they "know how to stride the essential and hold on to the root" (*Hanshu*, "Yiwenzhi"). Several other passages enhance this point:

> Now it is only oneness that is used to hasten transformation, and, by means of a few, is used to know the many. . . . He must be an upright person! For then he is able to grasp hold of correction to correct the incorrect, to lay hold of oneness and use it to know the many; to expel what is harmful to people and support

what is appropriate for them. In the total collectivity, he preserves oneness and possesses the same ends as heaven and earth: then he can know their calamities and good fortune. (*Shida jing*, ch. 9; Yates 1997, 137)

Now be oneness and do not transform,
Attain the root of Dao.
Grasp the few to know the many,
Attain the essentials of affairs.
Lay hold of the correct to correct the irregular.
First know highest antiquity,
Later [can] be quintessentially bright.
Embrace Dao and hold on to measures,
And the world can be unified. (*Daoyuan*; Yates 1997, 177)

Therefore, when there is oneness, there is nothing that cannot come into existence under heaven. If there is no oneness, there is nothing that can exist . . . "If you are able to hold on to oneness, the hundred beings will not be lost. If you cannot hold on to oneness, the hundred beings will be lost. (*Fanwu liuxing*)

Universal Reach

The reason why Huang-Lao thinkers are so keen on oneness as the key principle in governing the many is that they believe that oneness pervades and universally applies everywhere in the world and all through the people. Using their terminology, the sage king only needs to hold on fast to this ubiquitous oneness and, without ever having to set foot outside his dwelling, can perfectly pervade heaven and earth and make informed and beneficial decisions on all affairs even if they are a thousand miles away or anywhere within the four seas. This is made clear in the *Shida jing* dialogue between the Yellow Emperor and Li Mo:

In ancient times, when heaven and earth were first complete, the names were correct, the forms were harmonized, and all guarded the name of oneness. Above it was linked with heaven; below, it was applied to all within the four seas. I have heard about the complete laws of the world. Therefore, it is said, Do not use many, just speak one phrase and then stop. Follow names and go back to oneness, and the people will not disorder the rules . . .

In ancient times, august heaven caused the phoenix wind to descend to say one phrase and then stop. The five emperors used it to split apart heaven and earth, to administer all within the four seas, to cherish the people below, and to correct the officers of their generation. . . .

After it was lost, no one could guard oneness any longer. To understand oneness, examine heaven and earth. For the principle of oneness, apply it to all within the four seas. (ch. 9; see Yates 1997, 135)

The *Guanzi* echoes this, "For this reason, the sage comprehends it all in one phrase: above he examines heaven, below he inspects earth" ("Xinshu xia"). The *Huainanzi* says the same, "For this reason, he pervades the four seas. Based on his comprehension of oneness, he borders on heaven and earth" (1.4). The Mawangdui manuscripts use similar terms often interchangeably, such as "examine" (*cha* 察) and "border" (*ji* 際), the latter meaning something like "connect" or "reach to."[9]

The *Fanwu liuxing*, too, believes that Dao and oneness are universal and expresses this in terms very much like the above statements.

Attaining and fully comprehending oneness, he serves heaven above and nurtures all in the abyss below. Reflecting on things without ever leaving his seat, his plans reach as far as a thousand miles. Going about and applying it, he extends its benefits throughout the four seas.

Attain oneness and plan with it: align yourself with the world and assist it. Attain oneness and reflect through it: align yourself with the world and order it. Guarding oneness as the prime decree of heaven and earth.

Universally applying oneness and remaining steeped in it, therefore, the sage maintains it steadily, able to attain levels of strength and insight far beyond ordinary people. The *Fanwu liuxing* gives three examples:

Without using his eyes, he knows names;
Without using his ears, he knows sounds.

Thus I have heard: Holding on to Dao, he sits and never leaves the mat. Upright in his cap, he works with documents but never involves himself in affairs. He has precognition of all that occurs within the four seas, can hear as far as a thousand miles and see as far as a hundred. Therefore, the sage stays in his own place yet is the first to know whether his state or family is safe or in danger, will survive or perish, as well as all nefarious activities of robbers and brigands.

Grasp it and it is less than a handful;
Spread it out and it cannot be contained.

---

[9] Wang Bi, commenting on Laozi's expression, "embrace oneness," cites an *Yijing* commentary, saying, "Now, the masses cannot govern the masses. The only one who can do so is the solitary one. For this reason, in order to make the masses continue to exist one must most importantly connect to oneness."

Enlarge it and it can rule the world;
Reduce it and it can order the state.

The word translated "works with documents" (*zhu* 箸) in the second example is difficult to read, and scholars differ in their interpretation. Quite obviously, the term "upright in his cap" reflects the traditional garb of government officials, closely connected to the expression, "let fall the robes and face south," indicating an attitude of nonaction, clarity, and stillness. This means, the sage easily works with documents as part of his governmental function. Another possibility is that the term stands for "set forth" or "manifest" (*zhu* 著), meaning that sage is just present, remains clear and pure, and does not plan ordinary undertakings. Whichever it is, we can take this one step further. The sage in this vision has no need to leave his place to gather information: he knows, hears, and sees all from a distance, because he holds on to Dao, which pervades all.

Now the question arises: Why this pervasiveness? Where does oneness attain its universality? What makes it so ubiquitous? The fact is, it is because oneness or Dao resides deep within all beings and forms part of their original nature. In Huang-Lao thought, oneness creates the myriad beings in their innumerable forms; it is the key agent enabling the division of original unity into multiplicity, the factor that allows them to "attain" (*de*) or have virtue (*de*), the inner power that makes them what they are and connects them to Dao. From the perspective of the common "virtue" that each of the myriad beings possesses, what they each "attain" is oneness. The reason why oneness functions universally in all of the myriad beings is that, in fact, the commonality and pervasiveness of oneness resides deep within each being.

The *Zhuangzi* notes that philosophers Peng Meng, Tian Pian, and Shen Dao tended to focus primarily on "making all beings equal" (ch 33); the *Lüshi chunqiu* similarly says that Tian Pian "valued equalizing" (17.7). Equalizing the myriad beings means to see their inherent sameness, their fundamentally common nature, which in due course makes it possible to "spread equalizing and develop oneness." The myriad beings come in large numbers of different species and each have their unique individual differences. When Huang-Lao thinkers speak of equalizing them or recognizing their inherent oneness, they do not mean to disregard or denigrate their physical and phenomenal differences, but to discover their unity and common nature precisely in their variety, to open a vision of unifying standards and encompassing regulations in their multi-faceted so-being (see Mawangdui 1974, 72; Wang 2002).

The *Lüshi chunqiu* states, "Now, one who can make the differences among the myriad beings equal so that all ignorance and wisdom are leveled of and, to the best of his abilities exhausts his efforts to do so—seeing them all as if issuing from a single opening—this truly is a sage" (17.7). Huang-Lao thinkers do not try to alter the fundamental expectation that beings are different nor do they have any desire to see the dull and stupid as crafty and wise. They certainly accept that there is no way of changing the variegated nature of beings, yet they also insist that all beings have in common an equal urge to live in true accordance with their inherent nature, personal reality, and specific abilities, i.e., that they all issued from the same opening. Doing so, all beings share in cosmic oneness.[10]

The most important aspect of this oneness, moreover, is the overarching human urge to "pursue benefit and avoid harm." As the *Guanzi* puts it, "As for people's feelings and emotions, there are none who do not desire life and hate death; none that do not desire benefit and hate harm" ("Xingshi"). The *Shenzi* 慎子 expresses the same idea, using the concept of self-acting:

> Among people, there is none who is not self-acting. Yet if all were just self-serving, none could be employed. For this reason, the former kings did not appoint as ministers those who would not receive emoluments, did not give to those who were not wealthy already. If people cannot get what they need to be self-acting, their superiors should not employ them. For this reason, if you employ people who are self-acting and let those go who are self-serving, there will be none you cannot use. ("Neipian")

Since people individually have these kinds of urges and feelings or emotions, everyone must have them. They are common to all, which means that each person's feelings are in essence the same as those of all others, providing a basic equality among living beings. Huang-Lao thinkers, following this kind of logic, accordingly propose that in all sorts of multiplicity there is a unity, an inherent oneness which can be perceived by any organism in the world. This also means that one can understand other beings and places on the basis of one's own existence and locality. As Laozi says of the sage,

> Without ever stepping out the door, he knows the world;
> Without ever looking out the window, he sees the Dao of heaven. . . .
> Therefore, the sage
> Knows things without traveling,
> Names things without examining;
> Accomplishes things without acting. (ch. 47)

---

[10] The *Huainanzi* speaks of "making ordinary people" and "all beings equal" (ch. 11). This may well be due to the influence of the three Huang-Lao thinkers.

Wang Bi in his commentary to chapters 47 and 54 similarly says that "affairs have their ancestor and living beings have their leader," "all paths, however different, return to the same place," and "the hundred thoughts share one center," giving expression to the same idea from different angles. Why exactly a person should be able to fully grasp the "ancestor" or "leader" of the myriad beings and all things in the world without ever stepping outside his door, how exactly all things return to the same place, however, he does not clarify. Still, when commenting on chapter 54, "How do I know that the world is like this? By this," he says, and explains, "This asks the question what should I do to attain full knowledge of the world. The best way is to examine oneself and not look toward things on the outside. That is what Laozi means when he says, 'Without ever stepping out the door, he knows the world.'"[11]

The *Huainanzi* says, "A ruler of men need not descend from his palace halls to know about matters beyond the four seas, because he goes along with beings and to know about beings, he goes along with people to know about people. For this reason, where collective strength is employed, it is always victorious; where collective wisdom is employed, it is always successful (9.11; Major et al. 2010, 305). Matching this logic, the *Fanwu liuxing* says, "If you desire to hold on to oneness, raise your head and look at things, bow your head and examine things. Don't seek far off, but measure all within yourself." The latter phrase clearly speaks to the importance of examining oneself and understanding all from within as opposed to without. Knowing oneself is thus the core factor of all measures and guidelines of the world. Thus the cosmic logic of the one and the many turns into the structural patterning of the oneness and multiplicity of the myriad beings.

---

[11]The same passage is also explained in the *Hanshi waizhuan*: "In antiquity, they understood the world without going out of doors. They perceived the way of heaven without looking out of their windows. This was not because their eyes could see thousand miles ahead, nor because their ears could hear a thousand miles away, but because they measures others by their own feelings. From their own dislike of hunger and cold, the understood the world's desire for food and clothing. From their own dislike of toil and suffering, they understood the world's desire for peace and ease. From their own dislike of decay and poverty, they understood the world's desire for riches and abundance. Knowing these three things was how the sage kings put the world in order without ever descending from their mats [throne]. Hence the Dao of the ruler restes on sincerity and compassion" (ch. 3; Hightower 1952, 123)

Conclusion

In this chapter, I have examined the thought presented in the *Fanwu liuxing* in the light of Huang-Lao Daoism. Although a short text of only about 800 characters,[12] it presents a fundamentally important vision of oneness as pervading the universe, nature, and society. It clearly fits into the overall structure of the Huang-Lao school, occupying its own particular position within its history and lineages—the latter proposing different metaphysics while all trying to apply similar key concepts and foundational notions in their effort of understanding the world and its phenomena.

A central factor in all Huang-Lao and early Daoist thinking is the notion of oneness—also called Great Oneness or Dao—as representative of the highest ultimate of the universe and most fundamental principle of the world. One is not only the original number that gives meaning to all objects, but also the original cosmic state at the beginning of time. And just as "one" is closely related to all other numbers and the many, so original oneness remains forever connected to the multiplicity of the world and the myriad beings. This understanding in due course leads to the apperception of oneness as a central metaphysical concept, a highly abstract notion, an absolute value that provides an inherent commonality and universality among the unlimited multitude of existence. The relation of the one to the many, then, is the core model of Huang-Lao thinking in its effort at comprehending the world and its beings.

I personally think the concept of oneness is the most important aspect of the *Fanwu liuxing* as a Huang-Lao manuscript. Describing and analyzing the four different dimensions of oneness the text presents allows us to appreciate its function in more detail. There is no question that the text speaks directly neither of Laozi nor of the Yellow Emperor. If that were the main criterion for judging a text as part of the school, it would not belong here. On the other hand, if our main criterion is the philosophical character and the choice of key concepts, then, yes, the *Fanwu liuxing* most definitely is part of this tradition. Huang-Lao thinking consists of core *Laozi* concepts in conjunction with Legalist notions plus a certain influence of Confucian, Dialectician, and Yin-Yang philosophy. It is a complex system, syncretistic and integrative. As Sima Qian notes, characterizing Daoists among the Six Schools, they "go along with the great flow of yin and yang while adapting good

[12] The original editors of the manuscript supplied a few words to make up gaps, bringing the total to 846. If we leave out the passages from # 27, the total is 822.

points from Confucians and Mohists and observing the essential points of the Dialecticians and Legalists."

Accepting the Huang-Lao school as one branch of early Daoism after Laozi—Liezi, Yang Zhu, and Zhuangzi representing several others—the vexing question remains where exactly the *Fanwu liu-xing* fits into the overall picture and, more specifically, when it was compiled. Huang-Lao gradually developed and flourished in the Warring States period, rising to political prominence under the Han. Assuming that it began around 390 BCE, in the mid Warring States and grow until 140, when Emperor Wu ascended the throne, it was prominent for two and a half centuries.

Looking at the actual bamboo slips of the manuscript, they are, as Ma Chengyuan points out, covered in Chu-style writing and, on the basis of Carbon-14 dating as executed by the Shanghai Nuclear Center of the Chinese Science Institute, can be placed to 2257 years before the present, plus/minus 65 years (Zhu 2002, 3). That is to say, they were buried in the tomb of a local Chu lord around the year 278 BCE (Ma 2001b, 2). This provides the *terminus post quem non*, the latest point of compilation. The text was thus created before then, but the question remains, how far?

The reigns of the Qi kings Wei (356-320) and Xuan (319-301), i.e., the latter half of the 4th century, were an important period in the development of Huang-Lao. As Sima Qian says in the *Shiji*, in the biography of Mencius and Xunzi, about the thinkers Shen Dao, Huan Yuan 环渊, Tian Pian, and Jiezi 接子, "They all studied the arts of Huang-Lao and Dao-Virtue to discover their central ideas" (ch. 74). All these men, moreover, were closely associated with the Jixia Academy in the capital of Qi. The *Zhuangzi* classifies Peng Meng, Tian Pian, and Shen Dao as one school, specifying that Tian studied with Peng. The *Shiji*, in the biography of Laozi and Han Fei 韩非(ch. 63), adds that Shen Buhai 申不害 was "a student of Huang-Lao who focused on names and punishments." According to Zhang Dainian's dating of ancient thinkers, they all lived around the 4th century: Shen Buhai, ca. 401-337; Huan Yuan, ca. 390-300; Peng Meng, ca. 382-300; Tian Pian, ca. 370-290; and Shen Dao, ca. 370-290 (Zhang 1982, 28). Especially the last four lived under the reign of the two Qi kings who were major sponsors of the Jixia Academy. It is more than likely that they were involved, if not in the compilation, then in the transmission of the *Fanwu liuxing*, and that the text dates from this overall period or slightly earlier—from when exactly remains a mystery for now.

Chapter Four

Huangdi sijing

Governing through Oneness

The *Huangdi sijing*, unearthed at Mawangdui, presents Daoist thought of the Huang-Lao school. It focuses especially on how to create political and social order by "holding on to oneness" (*zhiyi* 执一) or "acting in oneness" (*weiyi* 为一). Most scholars to date have focused on the text's concepts of Dao and holding on to Dao (e.g., Mawangdui 1976; Wei 2004; Chen 2007a), but few have examined what it means by oneness. Both Dao and oneness are technical terms the text uses to indicate the original nature of the universe, and while it is perfectly legitimate to focus one's studies on Dao, a case can be made—however surprising it may seem at first glance—for a more in-depth analysis of oneness.

One reason is the particular emphasis the text places on oneness in conjunction with its political thought. Among all the various texts of the Huang-Lao school, the *Huangdi sijing* uses "oneness" the most, often interchangeably with "Dao," but also more specifically in contrast to the "many" (*duo* 多) or as representing the "singular" (*gongyi* 共一) in opposition to the "masses"(*wanchong* 万众), i.e., the ruling versus the ruled. Thus, I use the concept of holding on to oneness as the key to unlocking the political thought of the text.

Both Daoism in general and Huang-Lao in particular tend to use oneness and Dao interchangeably, so that expressions like "embracing oneness" (*baoyi* 抱一) or "guarding oneness" (*shouyi* 守一) are the equivalent of embracing or guarding Dao. They also both apply them to political thought and expositions on the arts of rulership—the latter reaching into profound philosophical dimensions and going far beyond the narrow sense of concrete, administrative measures or even the idea of "law" (*fa* 法) and "propensity" (*shi* 势) in the Legalist sense. The notion of "arts," moreover, implies the complex areas of power play and intrigue. It gives theoretical expression to general principles and rules of governance.

Rulership

The political principles of the *Huangdi sijing* also appear in the work of Sima Tan 司马谈, notably in his discussion of the Six Schools, as well as in the literary section of the *Hanshu*. For Sima, followers of Dao-Virtue "go along with the great flow of yin and yang while adopting good points from Confucians and Mohists and observing the essential points of the Dialecticians and Legalists." It also means "rooting oneself in emptiness and nonbeing and going along with it in all one's functions," leading to a state where one can "maintain alliances and go easy on control, have few undertakings and gain numerous merits" (*Shiji*, preface).

According to the *Hanshu*, the Daoist school describes "Dao of the past and today, of success and failure, arising and passing, good fortune and bad to the effect that one can know how to stick to essentials and hold on to the root, guard oneself in clarity and emptiness, and keep oneself safe in humility and weakness." This, then, respresents "the arts of the ruler facing south," a vision significantly different from Zhuangzi's "letting go and floating at ease." Generally Han authors, including Sima Qian, associate these arts or principles of rulership with the Huang-Lao school. I, too, see them as such, which also means that political thought of the period is strongly connected to it. Some people may ask: How come that certain materials that never even mention Laozi or the Yellow Emperor are still classified as Huang-Lao? The answer is that in classifying we look more to the philosophical content and lineage placement of a text than to its mention of particular figures. The school, therefore includes the concepts of Dao-Virtue as described by Sima Qian and the Daoism of the "ruler facing south" of the *Hanshu*.

The question then is, What is the relationship between Sima Tan's conceptualization of pre-Qin Daoism and what Sima Qian calls the "arts," "words," or "concepts" of Huang-Lao? As Qiu Xigui points out (1996), Sima's conception of "Daoist" is really quite narrow: he does not even include Zhuangzi and his followers.

A wider definition would involve certain aspects of Confucianism and Mohism and allow for a variety of different strands and lineages.[1] When Sima Qian, a generation later, speaks of Huang-Lao thinkers, he means people like Shen Dao, Tian Pian, Jiezi, Huan Yuan, and others, noting that they "all

---

[1] Han scholars define Huang-Lao as being short for the Yellow Emperor and Laozi. As Wang Chong 王充 says in his *Lunheng* 论衡, "Those pure in wisdom are Yellow and Old [Lao]. Yellow means the Yellow Emperor; Old refers to Laozi" (ch. 54).

pursue the arts of Huang-Lao and Dao and virtue" (*Shiji* 74). He similarly says that Legalists, like Shen Buhai and Han Fei, are rooted in Huang-Lao and base their thinking on it. This Huang-Lao, then, reflects the wider definition of pre-Qin Daoism, a tendency that probably goes back to the mid-Warring States period, when thinkers first began to combine Laozi's philosophy with Legalist and other concepts, such as the cosmology of yin and yang. When Sima Qian says that certain thinkers "are rooted in Huang-Lao," he certainly does not mean that they base their ideas on the Yellow Emperor and Laozi specifically, but that some of their principles and philosophical notions relate to ideas associated with them.

The *Huangdi sijing* similarly, while making direct reference to the Yellow Emperor, does not rely on Laozi's thought or cite his text and is yet firmly placed in the Huang-Lao camp (see Tang 1975; Xu 1989). Indeed, its theoretical outlook and essential philosophical concepts strongly reflect what Sima Tan and the *Hanshu* define as Huang-Lao. The last paragraph of the *Daoyuan* shows just how close its thought is to Huang-Lao arts of rulership:

> Now, act in oneness and do not change,
> Gain the root of Dao.
> Grasp the few to know the many,
> Gain the essentials of affairs.
> Lay hold on the correct to correct the irregular.
> Before, if you know all about high antiquity,
> Later you [can] know the quintessentially bright.
> Embrace Dao and hold on to the measures,
> And the world can be unified.
> Observe high antiquity,
> And discern everywhere it can be used.
> Inquire into matters before they turn to nonbeing:
> Get them when you can. (Yates 1997, 177)

Several terms here, such as "acting in oneness," "embracing Dao," and "holding on to standards," places the text clearly in the range of Huang-Lao arts of rulership. The passage is key to our appreciation of the entire corpus of four texts as reflecting rulership by holding on to oneness, however, it is not all that clear-cut. In the earliest redactions of the manuscript, it was punctuated differently, placing the first line of this passage separate and linking the later lines in a more purposeful way.

> Now, act in oneness and do not change.
> To gain the root of Dao,
> Grasp the few and know the many.

> To gain the essentials of affairs
> Lay hold on the correct to correct the irregular

However, this reading is not accurate. From the later parts of the passage, it is evident that the very first line cannot stand alone, but must connect to the second. Following this, the natural way of reading the text is as rendered first, which means that the inherent intention of the work provides clues to the way it should be read, and only careful analysis of parallel structures and matching words makes a correct understanding possible. It is thus quite clear that the first part of the passage addresses how the ruler should act while its later statements deal with the effects of his actions.

To paraphrase its content, an enlightened ruler, if he systematically follows the principles of oneness and never wavers, will get to the foundation of Dao. If he grasps even the minutest of things and uses this expertise to understand the inherent disposition of the myriad beings, he will master the essentials of political affairs.

Moreover, if he follows the proper way (rules and methods) and corrects all things disorderly and irregular, he can reflect on the past to observe reasons for rise and fall in ancient history as well as clearly see trends and dynamics of the future. If he firmly upholds Dao and observes all laws and standards, he will realize his desires and bring overreaching harmony and unity to the world. If he looks back to examine the distant past, he can fully grasp the dynamics of rise and fall in history. If he examines the trends and prospects of the future, he can attain accurate conceptions and positive guidelines.

The *Huangdi sijing*, when speaking of rulership by "acting in oneness," also uses expressions like "holding on to" or "guarding" oneness. For example, the *Shida jing* has, "Be centered in stillness without concerns, hold on to oneness without pursuing anything" (ch. 14), "Always guard oneness, and you will join the ultimate of heaven and earth; then you can know their good and bad fortune" (ch. 9), while the *Jingfa* notes that "heaven holds on to oneness" (ch. 6). What, then, does the text mean by oneness in concrete terms? Why is holding on to it sufficient to be a good ruler? How exactly does it help the ruler to realize his goals and expectations?

Dimensions of Oneness

The *Huangdi sijing* describes oneness in different dimensions: in its abstract understanding, it is Dao; in its concrete application, it is Dao's original state of chaos. Oneness is thus both a name for Dao and a way of referring to its origin. As the title of its fourth text, *Daoyuan*, suggests, it probes the "source of Dao," first by looking at its initial cosmic state. It says, "Empty and same, all was in oneness, constant oneness, nothing more." Oneness here indicates the merged and unified state of Dao at the beginning of creation. This state is also called Great Oneness; it comes before the division into multiplicity, before concrete forms or images emerge, and prior to names and appellations for things come about. "At the initiation of constant nonbeing, all was merged in great emptiness. "

The *Shida jing* describes it as "murky and vague, . . . making a single granary; without darkness, without light, prior to yin and yang" (ch. 2). These expressions are reminiscent of the *Laozi*, which describes Dao as "in chaos yet one" (ch. 25). Dao in chaos or oneness before the emergence of division rests in a state of constancy. However, one day it starts to separate and gives rise to the multiplicity of different things and beings. As the *Daoyuan* says,

> Now, it first splits to form the two,
> Divides into yin and yang,
> Separates to make the four seasons . . .
> It takes oneness as its measure and never changes,
> And is right there even for crawling insects.
>
> Birds attain it and fly.
> Fish attain it and swim.
> Wild animals get in and run.
> The myriad beings attain it and live.
> The hundred affairs attain it and are complete.
> (see Yates 1997, 173)

In addition, the text also uses oneness as an appellation for Dao: "Oneness, that's its name." Laozi similarly takes Dao as the root source of the myriad beings, but then says, "If forced to give it a name, I would call it great" (ch. 25). Oneness as an appellation for Dao implies certain content and qualities, which is similar but not the same as Laozi's use of the word "great." The *Huangdi sijing* using oneness to express the foundational state of Dao makes that clear. The *Shida jing* contains a number of dialogues between the Yellow Emperor and Li Mo. One begins with the question of whether oneness has content, to which Li Mo replies, "Oneness being the root of Dao, how can it not have content?"

The Yellow Emperor asked, "Is oneness just one and that is all, or does it also grow?"

Li Mo replied: "Oneness is the root of Dao. How could it be that it would not grow? . . . To understand oneness, examine heaven and earth. For the principle of oneness, apply it to all within the four seas. How can one know the beginning of an endless piece of string and the end point between far and near? Now it only oneness that is used to hasten transformation and, by means of a few, is used to know the many.

"Now, if you look thoroughly at the four seas to the farthest limit of up and down, and to the four directions that embrace each other, you will see that each uses its own Dao.

"Now, a hundred words have a basis, a thousand words have their essentials, and ten thousand [words] have their generalities. The multitude of the myriad beings all pass through a single hole. . . .

"Hold on to oneness to know the many; expel what is harmful to the people and support what is appropriate for them. Always guarding oneness, he possesses the same ends as heaven and earth: then he can know their calamities and good fortune." (ch. 9; Yates 1997, 135-37; Chang and Feng 1999, 169-70)

Here the text explains oneness from a variety of angles, notably in conjunction with qualities of heaven and earth and the myriad beings that are inherently universal. Oneness in this respect has two dimensions. It pervades both above and below, easily discerned by looking up to heaven and earth; it also is the gateway through which the myriad beings emerge in their multiplicity. The text here addresses the issue of oneness directly and clearly points out its quality as part of the original nature of the universe and its overarching universality. It shares the same syntactic role and characteristics as Dao.

Confucians often describe heaven and earth as the root source of the myriad beings; Daoists, in contrast, speak of oneness or Dao in this context, regarding heaven and earth as two kinds of beings, albeit extremely large ones. However, there are also numerous passages, where Daoists single out heaven and earth out from among the myriad beings and establish them as a model, thus giving them a special position. The *Zhuangzi* says,

If we calculate the number of beings that exist, the count certainly does not stop at ten thousand. Yet we set a limit and speak of the "ten thousand [myriad] beings" because we select a number that is large and agree to apply it to them. In the same way, heaven and earth are forms that are large, yin and yang are energies that are large, and Dao is the generality that embraces them. (ch. 25; Watson 1968, 291)

The *Huangdi sijing* would agree with that. It emphasizes oneness as the original state of heaven and earth but at the same time it also recognizes that heaven and earth hold a position of governance over the myriad beings. As the *Daoyuan* has it,

> Heaven and earth, yin and yang,
> The four seasons, the sun and the moon,
> The planets and constellations, primordial energy,
> The wrigglers that walk and crawlers that move,
> Even the plants that grow roots—
>
> They all take life from Dao,
> But never decrease it.
> They all return to Dao,
> But never increase it.
> (Yates 1997, 173-75; Chang and Feng 1999, 197-98)

Yet at times, the text also insists that heaven and earth have decision-making powers over the myriad beings. As the *Shida jing* has it,

> Those that followed heaven flourished;
> Those that opposed heaven perished.
> If they had not opposed the Dao of heaven,
> They would not have lost their guarded [possessions].
>
> Once heaven and earth were complete,
> The black-haired folk were born. (ch. 6; Yates 1997, 123)

Naturally, oneness or Dao ultimately form the root origin and foundation of the inherent universality of the myriad beings.

Another dimension of the *Huangdi sijing* in this context is that it connects oneness as root source and universal quality with the notion of principle (*li* 理). As is well known, the term does not appear in the *Laozi*, but it occurs variously in the *Zhuangzi* and rises to prominence in the *Hanfeizi*, where its relationship to Dao is defined in some detail. The two latter texts both use "principle" to refer to an underlying basis or organizational structure that all things and beings equally possess, yet which makes them different and unique. As the *Zhuangzi* says, "The myriad things differ in principle, but Dao shows no partiality among them, and therefore they may achieve namelessness" (ch. 25).

The *Huangdi sijing* speaks of principle quite often. It generally uses the term to refer to the particular type and specific structure of each of the myriad beings.[2] Thus, the *Jingfa* has, "When beings are in harmony wthrith Dao, we speak of principle. When some beings are not in harmony with Dao, we speak of losing principle" (ch. 6). Here "principle" designates the particular way of being of an individual entity. Since the text speaks of "holding on to" or "being straight with Dao and going along with principle," the term here can hardly refer to a foundational concept and must indicate the unique organizational structure of beings.

The *Jingfa* continues, "Going with or against are both part of Dao, but they depend on differences in principle." This clearly indicates something that varies among individuals. Let us look at some additional passages. For example, it also says, "He who holds on to Dao and grasps principle, must begin from the basis, acting by going along with the constant standards, prohibit and punish those guilty of crimes, always centered on the principle of heaven" (ch. 5; Yates 1997, 75). Here, "going along" means following the principle of heaven, which is a very similar notion as "principle of earth," Dao of heaven," or "Dao of earth," indicating the organizational structures and inherent patterns of different major entities. The *Jingfa* similarly notes,

> For this reason, the way one who holds on to Dao observes the world is that he always fully investigates and observes where affairs begin to arise, and also always investigates their forms and names. When forms and names have been fixed, going along and being contrary have their positions, the dying and living have their distinctions, and survival and destruction, as well as rise and decline, all have their places. After that, when he compares then with the constant Dao of heaven and earth, he determines the exact location of calamity and good fortune, death and life, survival and destruction, rise and decline. (ch. 8; Yates 1997, 95-97; Chang and Feng 1999, 140)

> Right and wrong have their appointed lots: decide them according to the law. Carefully listen in emptiness and stillness: take the law as your tally. To fully investigate and examine the principle of names as well as their beginning and end, this is called 'inquiring into principle.' If you are only public-spirited and without private bias, see and know without delusion, then your knowledge will rise up and flourish.
>
> The way, therefore, in which one who holds on to Dao observes the world is to see the correct Dao and go along with principle, know how to enumerate the crooked and the straight, and be able to list all ends and beginnings. Thereby, he can go along with names and inquire into principle. (ch. 9; Yates 1997, 101; Chang and Feng 1999, 143)

---

[2] In some places, it also indicates original universality, as in, "The principle of oneness pervades the four seas."

When the text in these two passages speaks of principle and correct Dao, it clearly indicates the organizational structures and inherent patterns of beings—a multiplicity of patterns as opposed to the one underlying cosmic principle. The latter is more like oneness: it underlies the myriad beings in a fundamental and universal manner, leading to a state where "oneness unifies multiplicity" and "through oneness one can rule the many." This is principle as oneness, Dao as law, i.e., the concrete application of the principle of heaven and earth and of the myriad beings.

Practical Activation

The arts of rulership in the *Huangdi sijing* in the widest sense describe ways of using oneness or Dao to face heaven and earth and the myriad beings; in a narrower sense, they indicate ways of dealing with the people and social realities. The text, as much as the Huang-Lao school, defines them as ruling by holding on to oneness. Besides this term, the text also uses words like "grasping," "acting in," and "returning to" oneness, as well as "guarding the one name" and "holding on to Dao." For example, the *Jingfa* has, "One who holds on to Dao observes the world;" and, "Only one who holds on to Dao can understand the turnings of heaven above" (ch. 1). It also says, "Therefore, only one who holds on to Dao can rest in emptiness and stillness, being public-spirited and deeply correct . . . maintain Dao and always do the right thing" (ch. 9); and, "One who holds on to Dao goes along with principle" (ch. 5). In addition, the *Daoyuan* speaks of "embracing Dao and holding on to the measures." In other parts, the *Huangdi sijing* uses expressions like "using this Dao," "holding on to this Dao," "grasping the few," and "establishing correctness."

Several questions arise. Why does the text pose holding on to oneness as the core feature of government? Why it is the best possible way to rule? And how is it most effective?

Oneness in the text is the most original state of the world, the deepest foundation of the universe, the myriad beings, and humanity. If a ruler can grasp hold of this, he will naturally have the core of rulership in his hands; if he can apply it in the world, he will obviously effect the greatest transformations—"there is nothing that is not done."

The *Daoyuan* explains it quite well, emphasizing that the ruler should act in oneness, maintain the root of Dao, and through grasping the few know the many. Its first part outlines the cause of a particular attitude or action, while its latter part describes its effect. This echoes statements in the *Shida jing* about the rule of the sage king or "corrected person." For example,

> He is able to grasp hold of correction to correct the incorrect, to lay hold of oneness and use it to know the many; to exple what is harmful to people and support what is appropriate for them. In the total collectivity, he preserves oneness and possesses the same ends as heaven and earth: then he can know their calamities and good fortune. (ch. 9; Yates 1997, 137)

These statements show the close interconnection between the cause of holding on to oneness and the practical results in the world.

Another relevant passage is the dialogue between the Yellow Emperor and Li Mo in the *Shida jing*, which raises issues of how the ruler is to deal with the people and social realities. The Yellow Emperor asks whether there is such a thing as "complete law" (*chengfa* 成法), i.e., firmly established and generally applied regulations that keep the people on the straight and narrow. Li Mo replies that indeed such a system exists and that it consists of guarding the "one name," "one phrase," and "oneness." It says,

> In ancient times, when heaven and earth were just complete, names were correct, forms were harmonized . . . was used to guard the one name. Above it was linked with heaven; below it was applied all over the four seas. I have heard about the complete laws of the world. Therefore it is said, 'Do not use many, just speak one phrase and then stop. Follow names and go back to oneness, and the people will not disorder the standards" (ch. 9; Yates 1997, 135)

When the text here speaks of "one name" or "one phrase," it really refers to oneness. According to this, all a ruler needs to do is go along with the "name" to recover original oneness. Once he acts from that position, the people will find it impossible not to behave appropriately and act within the law.

However, this "oneness" is a highly abstract notion, not easy to understand and appreciate. The Yellow Emperor accordingly questions whether it actually exists and is assured that indeed it does. Li Mo then provides some examples from high antiquity.

> In ancient times, august heaven cause the phoenix wind to descend to say one phrase and then stop. The five emperors used it to split apart heaven and earth, to administer all within the four seas, to cherish the people below, and to correct the officers of their generation. Now, for this reason slanderers all withdrew and worthy men rose up all together, the five evils fled, and the cunning debaters stopped. They all went along with the names and returned to oneness; the people did not disorder the standards. (ch. 9; Yates 1997, 135)

This echoes concrete instructions for ruling by going along with Dao already outlined in the *Laozi*, but goes into a few more details. To be sure, the text inherits some of Laozi's major concepts, such as emptiness, stillness, namelessness, selflessness, being without words, and having no desires. It also uses the notion of nonaction—by far the most important rulership guideline in the Laozi, frequently (at least fifteen times) referred to—but to a much lesser degree. For example, the *Daoyuan* says, "nonaction is its element," while the *Shida jing* notes, "Affairs constantly self-manage; I rest in nonaction" (ch. 15), and the *Jingfa* has, "Be without tenacity, position, action, and partiality" (ch. 1),

Conceptually the biggest difference to the *Laozi* is that the *Huangdi sijing* not only places no emphasis on guidelines to behave with weakness and softness and to remain free from contention, but even actively rejects these attitudes. It may on occasion encourage rulers to cultivate humility and awareness of their own weaknesses, but overall it favors strong action, using hard and soft equally, and even supports contention. The following passages provide an example.

> He who makes the hard into the soft will live; he who makes the soft into the hard will be attacked. He who values the soft is fortunate; he who values the hard will be obliterated. (*Jingfa*, ch. 9; Yates 1997, 99)

> Were it to use the strong to subordinate the weak, what state would not be victorious? Were he to use the noble to subordinate the mean, what man would not gain his ends? If he used the worthy to subordinate the worthless [what affair] would not be ordered? (*Jingfa* , ch. 5; Yates 1997, 77)

> When the Great [king] Ting ruled the world, he was calm, gentle, correct, and still. He determined that softness and rhythm should be established first. He was tactful, modest, reverent, and frugal. Humble and restrained, he advocated softness, always putting himself behind and not in front.
>
> He embodied correctness and honesty, thereby being benevolent. Kind and generous, he loved other people. He was upright, correct, and brave, never daring to put himself ahead of others. He was equable, still, and never rash; holding on to oneness, he never pursued anything. He modeled himself on the female and brought forth endless softness.
>
> [Therefore, resting in stillness,] he had correct virtue. He loved virtue and was not contentious. He took his stand on not daring and practiced not being capable. In battle, he showed he was not daring; in understanding, he held on to being incapable. Guarding weakness at all times, he continuously confirmed it; Waiting for male tendencies to exhaust themselves, he followed in compliance. (*Shida jing*, ch. 14; Yates 1997, 149)

> [Strength gives birth to authority; authority] gives birth to kindness; kindness gives birth to correctness; correctness gives birth to stillness. (*Jingfa*, ch. 6)

> Practice the laws and follow [Dao], [that is acting with] male and female. When male and female seek each other, hard will join with soft. When soft and hard compete with each other, male and female will take form. Below they accord with earth; above they accord with heaven. (*Shida jing*, ch. 2; Yates 1997, 107)

> Now, to fight is inauspicious, but not to fight will also not complete the task of bringing good order to the world. The Yellow Emperor thereupon brought out his square-holed ax and his great ax and seized his weapons. He personally held the drumstick to meet Chi You and thus captured him. (*Shida jing*, ch. 3; Yates 1997, 115)

The positive attitude the text shows toward strength and contention marks, as Chen Guying suggests, a particular development in Daoist thought and constitutes the greatest change in Huang-Lao thought (2007, 404-05). It forms part of the active combination at the time of Laozi's Dao with law, which made it possible for the general guideline of ruling by holding on to oneness to turn into a systematic system of government. What, then, did that look like?

Social Dimensions

Combining Laozi's Dao with the concept of law, Huang-Lao thinkers understood the latter in two dimensions. For one, they saw it as the ultimate root cause of the universe, containing both an overarching universal nature as well as maximum legitimacy and effectiveness. For another, they saw holding on to oneness as key to its operability in the world. [3] The following statement from the *Jingfa* makes this clear:

---

[3] "Law of Dao" here can be read as "divine law." As the *Jingfa* (ch. 9) says, "Dao is the source of spirit and light. Spirit and light reside inside measures, but they are also visible outside them. They are inside because they are trustworthy even though they do not speak. They are visible outside because they do not change even though they may speak. They are inside also because they cannot be shifted even though they are still. They are visible outside also because they cannot be transformed even though they move. Still yet not shifted; moving yet not transformed: thus, we speak of spirit. Spirit and light are the epitome of perception and knowledge" (ch. 9; Yates 1997, 99).

> Dao produces law. Law is what draws the line between gain and loss, and makes clear the crooked and the straight. He who holds on to Dao, therefore, produced law and does not dare transgress it, establishes law and does not dare to violate it. [Therefore,] if he is able to align himself, he will not be confused when he sees and knows the world. Empty and formless, deeply serene and mysterious, the myriad beings all take birth from it. (ch. 1; Yates 1997, 51; Chang and Feng 1999, 100)

Law is Dao or oneness made concrete, embodied on the human plane, established through Dao or oneness as highest political authority. Law as universal force appears first as the abstract universal notion of rulership through holding on to oneness, then transforms into concrete ways of governance through holding on to the one law. In the *Laozi* there is no direct connection between law and embracing or holding on to oneness, but in Huang-Lao the sage activates it in his impact on the world.

Thus, the ruler can "draw the line between gain and loss, and make clear the crooked and the straight." The *Shida jing* has, "I have heard about the complete law of the world. Therefore, it is said, 'Do not use many, and just speak one phrase and then stop. Follow names and go back to oneness, and the people will not disorder the standards" (ch. 9; Yates 1997, 135). By the same token, the *Heguanzi* says, "That which guards oneness and thereby controls the myriad things is law" (ch. 8; Wells 2013, 134). It also notes,

> Oneness makes law to complete its enterprise:
> So that all will make it Dao.
> When the one law is established,
> the myriad beings all come in allegiance. (ch. 5; Wells 2013, 109)

The *Huangdi sijing* emphasizes the public and correct nature of law, which extends to the ruler as he maintains it by holding on to oneness.

> Laws and measures are the epitome of correctness. He who can govern by means of laws and measures will not suffer disorder. He who can promulgate laws and measures will not suffer disorder. To be quintessential, public-spirited, and without private bias, and to have reliable rewards and punishments is the way to govern. (*Jingfa*, ch. 3; Yates 1997, 63)

> If at rest a person possess laws, and in movement and action follows the correct names, his affairs will easily complete. However, if a person's affairs lack constancy, if he exceeds the limit and fails to correspond with what is appropriate, if he alters old [models] and changes the constant regulations, then he has no virtue. In managing punishments, he does not match them with the crimes; at rest, he lacks laws, and in movement and rest, he is not in harmony with the

> correct names. For this reason, he will suffer execution, receiving punishment for his actions. (*Shida jing*, ch. 6; Yates 1997, 125-27; Chang and Feng 1999, 163)

Besides holding on to oneness, the ruler here also has to honor all forms and names as well as go along with all social divisions and ranks (Cao 2010, 2-31). In the widest sense, these indicate the various rules and regulations of society. As the *Jingfa* describes it,

> The way of seeing and knowing merely means to rest in emptiness and nonbeing. As for emptiness and nonbeing, an autumn hair brings an object into existence, and then it inevitably has form and name. Once form and name are there, there is the division between black and white. Thus, someone holding on to Dao looks at the world without tenacity, is without firm position, and remains in nonaction and impartiality.
>
> For this reason, there is no affair in the world that does not appear in form and name, reputation and appellation. Once forms and names are established, reputation and appellation set up, then there is nowhere to conceal one's tracks or hide one's true aims.
>
> The public-spirited are bright; the highly enlightened are most successful. The most correct rest in stillness; those of utmost stillness are sages. Those without personal bias are wise; those of highest wisdom are models for the world.
>
> Measure them by the steelyard and align them with the correspondence to heaven. Material affairs must receive minute examination, for affairs are like straight trees, as numerous as grains in the granary. When the picul and pint measures have been provided, the foot and inch rules arranged, no item could escape its true spirit.
>
> Thus, we say, when rules and measures have been provided, order and control them. (ch. 1; Yates 1997, 51-53)

This closely echoes the *Shida jing* description of the ruler as "a person at rest who possess laws, and in movement and action follows the correct names. The term "names" here indicates the complex administrative system, including all official norms and permissions, that runs governmental life and political order and determines the social positions and ranks of each individual in the entire society.

Law, then, divides into the following major categories: [4]

---

[4] The *Shida jing* similarly says, "Heaven possesses constant stability; Earth possesses constant regularity. Share your affairs with the people; merge your [radiance] with the spirit. The arrogant and excessive love to fight and secretly plan inauspicious activity. They will be punished with strong injunctions and suffer the danger of death and extinction. . . .

> Heaven and earth have constant regularity; the myriad people have constant affairs. Noble and base have constant positions; there is a constant way in caring for ministers; there are constant rules for employing the people.
>
> The constant regularity of heaven and earth manifests in the four seasons, darkness and light, growth and decay, soft and hard.
>
> The constant affairs of the myriad people are that males till the soil while females weave.
>
> The constant positions of noble and base are that the worthy and worthless do not hinder each other.
>
> The constant way in caring for ministers involves employing the capable so that they do not exceed their duties.
>
> The constant rules for employing the people mean to remove private bias and establish public interest.
>
> Change and constancy, excess and abiding by the rules control each other by the incorrect. When correct and incorrect occupy their proper positions, names [and forms] will not leave them.
>
> In general, affairs do not divide into great and small, but beings set up dwellings of their own accord. Going along or against, dying or living, beings create names of their own accord. Once names and forms have been established, beings correct themselves of their own accord, (*Jingfa*, ch. 1; Yates 1997, 53-55)

This means that names and divisions are the concrete manifestation of oneness the ruler should hold on to. For the *Huangdi sijing*, a good ruler always strictly adheres to all forms and names: he only "distributes things according to their divisions" and "bestows things according to their names." Thereby he makes it so that "the host of the people do not contend" and "the myriad beings are settled by themselves" (*Daoyuan*).

Besides emphasizing the need to hold on to the universal law of Dao and all forms and names, the text also notes that rulers should honor and observe the principles inherent in affairs, beings, and nature. As outlined earlier, principle relates variously to oneness, but the text here takes it one step further, widening its range to include all kinds of order in nature and behavioral structures among humanity. It distinguishes heavenly from earthly and human principles: the ruler needs to honor them all:

"Heaven hates loftiness; Earth hates vastness; human beings hate harshness. Grow lofty without stopping, and Heaven will defeat you. Grow vast without ending, and Earth will abandon you. Be harsh without stopping, and people will kill you" (ch. 13; Yates 1997, 147; Chang and Feng 1999, 175).

> The sage does not create the beginning and does not monopolize himself; he does not make plans beforehand, nor act for gain or reject good fortune. Rather, he relies on the guidelines of heaven. (*Cheng*; Yates 1997, 157).

> When the initiates an affair, he harmonizes with heaven and earth, conforms to the people, and is favorable to ghosts and spirits. He causes the people to share their benefits and the myriad folks to rely on him. Thus, we speak of righteousness. (*Shida jing*, ch. 12; Yates 1997, 143)

> One who holds on to Dao and follows principle must begin from the basis, compliantly acct according to the constant standards, prohibit and punish those held guilty of crimes. He must center himself on the principle of heaven. (*Jingfa*, ch. 5; Yates 1997, 75)

> Generally, when someone violates the prohibitions and abandons principle, Heaven will punish him to the extreme. (*Jingfa*, ch. 7; Yates 1997, 89)

The prohibitions and taboos the text outlines are extremely broad, reaching through all different dimensions: nature, society, and individual.

> Palaces and buildings that exceed the measures are what the highest emperor hates. He who builds them will not dwell in them. Even if he manages to live in them, he will inevitably be exposed.
>
> To reduce the amount of clothing and coverlets, and diminish the widths of the outer and inner coffins: that's a prohibition. To have rapid or harsh corvée labor projects that could destroy marshland: that's a prohibition. To pick foliage and pile up wood for coffins that could inflict damage on forests: that's a prohibition. To gather [soil] to reduce heights and raise low ground: that's a prohibition. Only in times of a great flood are these actions permissible.
>
> Do not get ahead of heaven's [process of] completion; do not grow things out of season. If you get ahead of heaven, you will suffer destruction; if you grow things out of season, you will not attain results.
>
> The sun makes it bright; the moon makes it dark. Rest when dark; arise when light. Do not miss heaven's limits, but scrutinize the numbers and stop in time. (*Cheng*; Yates 1997, 161-63)

> Doing what goes against constancy: heaven prohibits it. Failing in one's duties: earth prohibits it. Disobeying orders: the ruler prohibits it. Once these three prohibitions are practiced, the state is close to perfection. The prohibitions of earth say: do not [to take away] from the lofty; do not add to the low. Do not block up rivers; do not go against agricultural needs. Do not oppose the people's bright intelligence. (*Shida jing*, ch. 10; Yates 1997, 139)

Laozi's rule by nonaction is mainly based on the notion that "Dao models itself by being just so" (ch. 25), working closely with public opinion and accepting individual intentions by encouraging the "sage to support the self-being of the myriad beings and never dare to act" (ch. 64), ideally by "making the mind of the people his own" (ch. 49). Huang-Lao thinkers developed this vision to include the important concepts of cause and honoring, that is to say, the ruler should find clear cause of action by honoring the special characteristics of all things and beings, and most importantly the wishes and interests of the people. "Go when you are going; stay put when you are put. Rely on the earth for goods; rely on the people for troops. Unless you rely on them you will not have a good basis for results" (*Cheng*; Yates 1997, 161).

According to Huang-Lao thinking, the wishes and actions of the people form part of their original tendencies and basic natures. Due cause of action for the ruler, in concrete terms, then means honoring and satisfying people's natural tendency toward personal self-interest. As Shenzi says when speaking about "self-acting" (*ziwei* 自为), i.e., people acting for themselves,

> Among people, there is none who is not acting for himself. Yet if all were just self-serving, none could be employed. For this reason, the former kings did not appoint as ministers those who would not receive a salary, did not give to those who were not wealthy already. If people cannot get what they need act on their own behalf, their superiors should not employ them. For this reason, if you employ people who act for themselves and let those go who are self-serving, there will be none you cannot use. ("Neipian")

The *Huangdi sijing* also uses the term, when the *Cheng* says,

> He who does not receive a salary, the son of heaven will not appoint as a minister. He whose salary is meager will not join others in committing transgressions and thus create difficulties for the ruler. Therefore, it is best to use those who act for themselves.

Just by going along with people's self-interests, the ruler can, therefore, sufficiently establish a world order of public uprightness and social stability. As the *Shida jing* has it,

> If you wish to know about gain and loss, you must essentially investigate names and examine forms. Forms constantly are self-determining, and for this reason, I am more at rest. Affairs are constantly self-managing, and for this reason, I remain in nonaction. Be still and hidden, and do not move. What comes will arrive by itself ; what goes will depart by itself.
>
> Can you be one? Can you stop? Can you be without a self? Can you choose yourself and still honor principle? . . .
>
> When the myriad beings flock together, I cannot but respond. (ch. 15; Yates 1997, 153)

Conclusion

In this section, we have examined the *Huangdi sijing* notion of rulership by holding on to oneness in various dimensions. First, we inquired what oneness means and how holding on to it forms the core of the arts or rulership, then we clarified the relationship of Han Daoism and Huang-Lao to understand the standards established by the school and its connection to the text. On this foundation, we moved on to examine what the *Huangdi sijing* specifically means by "oneness" how it sees the ruler applying it as a universal and widely effective measure. We also looked at how concrete rules and regulations combine oneness with the notion of law and how the universal activation of holding on to oneness transforms into the practical operation of rulership according to law in the text. Placing all this in a wider context, we saw how pre-Qin Daoism developed from Laozi's political philosophy into Huang-Lao doctrine and the thought expressed in the *Huangdi sijing*. We also understood how the abstract notions of oneness and Dao transformed into concrete legal and administrative systems.

Chapter Five

Laozi

Dao Models Itself

"Dao models itself on self-being" [*dao fa ziran* 道法自然] is a well-known yet frequently misread statement in Laozi's thought. Today contained in chapter 25, it also appears in the earliest known edition, the Guodian bamboo slips (divided into three versions A, B, and C), suggesting that it was one of the original themes of the work. The bamboo version reads:

> There is a something, in chaos yet complete,
> Ever since before heaven and earth were born.
> Silent it is, and solitary;
> Standing alone, it never changes.
> It can be considered the mother of the world.
>
> I do not know its name.
> To give it an appellation, I speak of Dao.
> Forced to give it a name, I call it Great.
> Great means it departs.
> Departing means it moves off.
> Moving off means it returns.
>
> Heaven is great, earth is great,
> Dao is great, the king is great.
> The domain has four great entities:
> The king rests as one of them.
>
> Humanity models itself on earth.
> Earth models itself on heaven.
> Heaven models itself on Dao.
> Dao models itself on self-being. (see Henricks 2000)[1]

---

[1] The Tang scholar Li Yue 李约 in his *Daode zhenjing xinzhu* 道德真经新注, punctuates the last section on the "four greats" differently: "The king [human being], modeling himself on Earth, is [becomes like] Earth; modeling himself on Heaven, is

All editions—Guodian, Mawangdui, and the transmitted version—consist fundamentally of the same words, with the exception of "something" (*zhuang* 状) in the first line, which in the transmitted version reads "a being" (*wu* 物), placing Dao somewhat outside of the scope of beings. Our focus here is how to interpret the last line, "Dao models itself on self-being."

Commentators and scholars for the most part have read this to mean that Dao models itself on itself or that it does not have anything further to model itself on, greatly influencing the perception of Laozi's thought. I myself started out thinking along the same lines, but over the years have come to feel increasingly doubtful.[2] The passage is not only tricky in terms of words and syntax but also philosophically different in the wider context of Laozi and early Daoist thought. Jiang Xichang (1988), Ikeda Tomohisa (1993), and Wang Bo (1995), in their studies of Laozi's concepts of self-being and nonaction, all made suggestions that have guided me toward a new perspective. My various inquiries into Laozi and early Daoist thought have focused increasingly on this specific passage. Below, I first outline the most common previous readings and their provenance, then reexamine the statement based on the various *Laozi* editions as well as in the light of the philosophical structure of Laozi's thought. I thereby hope to come closer to understanding what the passage really means.

Previous Readings

For the most part, contemporary *Laozi* scholars and commentators agree to read the passage to mean that "Dao it just as it is" or "has nothing further to model itself on." Feng Youlan, for example, notes that the term "self-being" here merely means that Dao gives birth to the myriad beings without any particular purpose or conscious cognition. He says,

> This does not mean that above and beyond Dao there is yet something else, which Dao would model itself on. The line above it reads, "In the domain, there

---

Heaven; modeling himself on Dao, is Dao: he models himself on self-being." He also claims that previous exegetes had misread the text and punctuated it wrong. Gao Heng thinks he has a valid point. "Li Yue's reading is clever and bright, very skillful. It makes me think that the original text was simply, 'The king models himself on Earth, on Heaven, on Dao, on self-being,' and that the repeated characters were added later" (1988, 61-62). The Guodian version has proved them wrong.

[2] This began in 1994, when I presented a study of the passage in the commentaries by Wang Bi, Heshang gong, and modern scholars at the Luyi Laozi Symposium.

> are four great things," referring specifically to humanity, earth, heaven, and Dao.
>
> The term "self-being" merely expresses that Dao gives birth to the myriad beings without any specific purpose in a process free from conscious cognition or intention. "Self-being" here functions as an adjective, it is not a separate entity or thing of any kind. Thus, the text speaks of "four" great things and not of "five." Laozi, when he says, "Dao models itself on self-being," stands in clear contrast to all teleological thinking. (2000, 254)

Zhang Dainian similarly says that the passage means that "Dao takes itself as its model" (1989, 79), while Ren Jiyu directly translates it as "Dao models itself on itself" (2006, 56). Tong Shuye sees self-being as a fundamental characteristic of Dao. He says, "The expression self-being in Laozi's book means being of itself and just itself. The passage, 'Dao models itself on self-being' points to self-being as an inherent, fundamental characteristic of Dao" (1982, 113). Along the same lines, Chen Guying emphasizes that Laozi's Dao modeling itself on self-being indicates that Dao purely goes along only with itself and has nothing else to model itself on" (2007, 173-74). He refers back to Wang Bi's reading of the passage:

> The passage "Dao models itself on self-being" has caused much consternation in readers. Dao in Laozi's thought is the ultimate concept, the force that originates and guides all existence. So, how could this Dao still need anything else to model itself after? In fact, already Wang Bi comments on the passage by saying that "Dao is no different from self-being." That is to say, the process and function of Dao is to flow along with self-being. (2007, 359)

Xu Kangsheng, moreover, bases his reading on the Han commentator Heshang gong 河上公, "Dao's nature has nothing to model itself on. In Heshang gong, self-being is not an entity separate from Dao, but just Dao being itself. The passage essentially means that Dao, as the first cause of heaven and earth, does not need anything else to model itself on, but exists solely by modeling itself on its own self-being" (1985, 114). Liu Xiaogan points out that the word "model" appears four times in the chapter and should be read in the same way each time. This would preclude a distinction between the first three and the last occurrence. While he insists on avoiding the grammatical inconsistency, he yet contends that self-being is essentially the inherent nature of Dao, its highest principle and deepest root (2006, 288-91).

Historically, too, we have plenty of readings.[3] The *Hanfeizi* in its two chapters on the *Laozi* (chs. 20, 21), the to-date earliest commentary on the text, unfortunately does not mention the passage. The literary section of the *Hanshu* lists three *Laozi* interpretations that have not survived, by scholars named Ling 邻, Fu 傅, and Xu 徐. The earliest extant Han commentary by Heshang gong, as noted earlier, sees the passage as indicating the basic nature of Dao as having nothing to model itself on (2.25ab).[4] Under its impact, many later *Laozi* exegetes have opted for a similar reading. Thus, for example, the Song scholar Lin Xiyi 林希逸 in his *Laozi kouyi* 老子口义 explains that Dao is self-sufficient.

> Dao is its own root and source, existing from before heaven and earth, stable and steady from high antiquity. It does not have anything else to model itself on. Without any model, it is its own self-being and nothing more. Thus, the text says, "Dao models itself on self-being."

Ge Changgeng 葛长庚 in his *Daode baozhang* 道德宝章 similarly takes the passage to mean that "Dao is in itself just as it is."

Wu Zheng 吴澄, in his commentary, agrees with this, adding that Dao is essentially nonbeing and namelessness.

> Humanity here indicates the sage. Modeling is following a certain standard and reaching to a particular level. Humanity is great because it attains Dao and reaches out to become one with earth: it models itself on earth. Earth is great because it attains Dao reaches out to become one with heaven: it models itself on heaven.
>
> Heaven is great because it becomes one with Dao: it models itself on Dao. Dao is great because it is just its own self-being. Thus, the text says, it models itself on self-being. There is no separate entity outside of Dao. Self-being is the quality of nonbeing and namelessness.

Jiao Hong 焦竑, in his *Laozi yi* 老子翼, accepts Lin Xiyi's understanding and sees self-being as an expression for not having anything to model itself on. He says, "Dao takes non-modeling as its way of modeling. Non-modeling is plain self-being. Thus, the text says, 'Dao models itself on self-being.'" Wei Yuan 魏源, in his *Laozi benyi* 老子本义, similarly says, "The self-being in the

---

[3] For a contemporary collection of *Laozi* commentaries, see Chen 1984; Ding et al., forthcoming. The latter presents a multi-year effort and will present well over 2000 *Laozi* editions and commentaries.

[4] The other Han reading, by Yan Zun 严遵, the *Laozi zhigui* 老子指归, does not comment on this passage. The Tang scholar Qiang Siqi 强思齐, in his *Daode zhenjing xuande zuanshu* 道德真经玄德纂疏, cites a commentary to the chapter, but only speaks of "merit and virtue being joined" (Wang 1994, 141).

passage, 'Dao models itself on self-being,' indicates inherent nature, more specifically the original inherent nature of Dao. Dao modeling itself on self-being, then, means that Dao originally is self-being and anything modeling itself on Dao in fact models itself on its self-being" (2.1). Seeing self-being as an expression of inherent nature, and as the particular nature of Dao, is unique to Wei Yuan. Xiong Jilian 熊季廉 reads "model" as "an expression for having a certain set framework and not going beyond this." Yan Fu 严复 acknowledges this when he says, "This really is a comment on the corruption of the text" (1983, 1085). For him, Dao is self-being, so that "Dao models itself on self-being" is just another expression of Dao modeling itself on itself.

There is no point in presenting further examples. The consensus among the majority of commentators is that the passage means that Dao simply models itself on itself, there being no further force or entity beyond it.

Issues of Interpretation

Some questions remain. For one, the last instance of the word "model" is quite superfluous. If, as the commentators say, the passage simply means that Dao in fact is self-being, there is no need to have the word "model" in it. In the preceding lines, each entity clearly models itself on something other, something higher and greater than itself: humanity→earth→heaven→Dao. If Dao does not model itself on something else, why have the same formulation? Also, if the word were left out, the pattern would reverse to a three-character line, matching the other verses. Commentators have tended to read the first three instances of modeling differently, singling out the phrase where Dao is the modeling agency as special. Philologically this is not sound practice.

Another issue that comes up is the reading of self-being as an aspect of Dao itself. Some commentators emphasize the meaning of "model" as involving an object, however, they still tend to see self-being as a characteristic of Dao, so that Dao again ends up modeling itself on itself. This is not really the case. They just come to this conclusion because they cannot imagine that there might possibly be something separate, higher, beyond Dao, since Dao is the root source of the myriad beings, the ultimate reality of the world.

The problem is that commentators never really examine how Laozi uses "self-being" in other chapters. They overlook the fact that the term appears most commonly in conjunction with the myriad beings and hundred clans and see it as an aspect of Dao. In fact, "Dao models itself on self-being" does not speak of an inherent characteristic and mode of functioning of Dao, but indicates the natural way of being of the myriad beings and hundred clans. The correct reading of the passage, then, is "Dao goes along with the self-being of the myriad beings," the way things and beings evolve naturally. In fact, Wang Bi already pointed in this direction. He reads "model" as "not go against," that is, as "go along with" and links it with the myriad beings. He says,

> "To model on" means "to follow the example of." By modeling itself on earth, humanity avoids acting contrary to earth and so attains perfect safety. By modeling itself on heaven, earth avoids acting contrary to heaven and so attains its capacity to uphold everything. By modeling itself on Dao, heaven avoids acting contrary to Dao and so attains its capacity to cover everything. By modeling itself on self-being, Dao avoids acting contrary to self-being and so realizes its own inherent nature.
>
> To model itself on self-being means that, when it exists in a square, it models itself on the square; if it exists in a circle, it models itself on the circle: it does nothing contrary to self-being. Self-being, moreover, is a term for that for which no equivalent exists, an expression for that which has infinite reach and scope. . . .
>
> Dao modeling itself on self-being results in heaven having something to rely on. Heaven modeling itself on Dao results in earth having something to emulate. Earth modeling itself on heaven results in humanity finding images in it. The reason why the king can be master of all is that he treats what he rules in oneness. (Lou 1980, 65; Lynn 1999, 96-97)

When Wang Bi says, "Dao does not go against self-being," he most certainly does not mean that Dao simply goes along with its own self-being or that it is just like this in itself. Rather, he means that Dao simply goes along with the self-being of the myriad beings, with the way they are of themselves. Modeling means to go along with; it is the same as not going against or following, and in all cases stands in close connection to the myriad beings. Wang Bi is very clear and concise on this point, but many later commentators followed the reading of Heshang gong instead, moving into an erroneous direction.[5]

When Wang Bi says that Dao does not go against self-being, he does not mean that Dao simply follows its own self-being or is just as it is, but points to Dao matching the self-being of the myriad beings. This reflects a core contention, a fundamental building block of Laozi's thought. Dao as ultimate reality is honored precisely because it gives birth to the myriad beings; this is its "mysterious virtue." It is amazing, enormous, limitless: not only can it give birth to

---

[5] The one exception is Lu Yusan who follows Wang Bi. "Here self-being does not indicate the natural world but means being in itself. The Dao originally does not do anything or bring anything forth. It just goes along with the self-being of the myriad beings, matching them closely. Because it flow with nature, it can bring forth and raise the myriad beings" (1987, 129).

the myriad beings but it can also encompass them in their complexity without ever manipulating or controlling them—like parents giving birth to children and raising them without overly controlling them, allowing them to unfold freely into whoever they are. Other images in the text are the ocean and the sage: the ocean is the highest reservoir and deepest resource of all the rivers and valleys, into which their goodness flows; the sage is greatest human being, the foundation of society, ubiquitous yet humble.

Self-Being in the World

The term "self-being" does not occur very often in the *Laozi.* The transmitted version only has five instances, three of which match the Guodian manuscripts. Aside from the passage discussed above, these are:

> Therefore,
> The sage desires what [others] do not,
> And does not value goods that are hard to get.
> He teaches what [others] do not
> And returns to what the multitude passed by.
> Thus, the sage can support the self-being of the myriad beings
> And never needs to act. (ch. 64)
>
> The work [of the early rulers] was done
> And their undertakings were successful.
> Still, the hundred clans said,
> "We do this by self-being!" (ch. 17)

The Mawangdui version has a variation in the last line, reading, "The hundred clans speak of themselves as self-being," which continues in the transmitted version, which also adds the word "all" after "the hundred clans ." The latter two also use the term in the following instances:

> Self-being has subtle words. (ch. 23)
>
> Dao gives birth to them;
> Virtue nurtures them;
> Beings give them form;
> Circumstances complete them.
> For this reason,
> There is none among the myriad beings
> That does not venerate Dao and value virtue
> The veneration of Dao and valuing of virtue

Is not due to some outside command,
But constantly self-being. (ch. 51) [6]

Among these five passages, two clearly indicate self-being as an aspect of the myriad beings or hundred clans. Chapter 17 insists that good rulership means that the people are just who they are by themselves, resting in their self-being and taking credit for social successes and accomplishments. Chapter 64 praises the sage as one who supports all beings in who they are yet never does anything actively to control or steer them. The quality of self-being is clearly attributed to the myriad beings or hundred clans. The *Hanfeizi* interprets the latter passage,

> Indeed, everything has a definite shape. It should be put to use accordingly. Thus, one should go along with its shape, which means, stand on virtue when in stillness and follow Dao when in motion.
>
> Once a man of Song made a mulberry leaf of ivory for his ruler. Having stems and branches, wide and narrow, as well as tiny buds and colorful gloss, it was scattered among real mulberry leaves and looked not different from them. As a reward for his skillfulness, the man was endowed with a bounty in the state of Song.
>
> Liezi heard about this and said, "If heaven and earth made one leaf every three years, leaves would be few indeed." Therefore, if you do not count on the natural resources of heaven and earth but look to one man for everything, of if you do not go along with the course of reason and principle but learn from the wisdom of one man, it is the same as making a single leaf in three years.
>
> Thus, farming in winter, even the divine Hou Ji would not be able to turn out good crops. On the other hand, rich harvests in years of abundance even bondmen cannot spoil. Thus, if you depend on the power of one man, even the divine Hou Ji would not be good enough, but if you follow the self-being of the universe, bondmen will be plenty. Thus, we say, "Cherish the self-being of the myriad beings, but never act upon them." (ch. 21; Liao 1959, 220)

When the text speaks of "beings," it refers to the myriad beings; when it speaks of "holding on to them," it means to "go along with" or "match" them, that is, move in accordance with their self-being. The usage of self-being here clearly is not as an aspect or characteristic of Dao or even of the sage, but indi-

---

[6] The Mawangdui version has "decree" for "order" and uses *heng* for "constantly" instead of *chang*. Jiang Xichang connects the phrase, "None among the people directs it, yet it is self-balanced" (ch. 32) with this, pointing out the parallel formulation. "Decree" and "order" as well as "self-balanced" and "self-being" are very close, as are "order" and "decree" (1988, 317).

cates the concrete tendencies and ways of life of the myriad beings and hundred clans.

Although not expressed as directly in the remaining three passages, the term self-being here, too, connects to beings rather than Dao. "Self-being has subtle words" literally means that by speaking little one can match the self-nature of things. By extension, it indicates that the ruler should issue few orders or decrees and instead match the self-being of his subjects. It resonates with chapter 5, "Many words lead to quick exhaustion—much better to guard it within." Many words is clearly the opposite of subtle or few words, a theme also present when Laozi speaks of the sage "practicing the teaching of no words" (ch. 2).

Keeping speech to a minimum is a requirement of good rulership: the fewer measures, decrees, and orders the ruler issues, the more freely the people can unfold their abilities and pursue their affairs. Jiang Xichang links it with the contrast of assertive action and nonaction,

> The term "words" in Laozi typically has to do with voicing teachings and issuing decrees. . . "Many words" means governing through lots of measures; "subtle words" indicates ruling with just a few. One matches assertive action; the other stands for nonaction. This is made clear in chapter 17. When chapter 23, moreover, says that "self-being has subtle words," it indicates the sage governing through nonaction, letting the people fulfill themselves. (1988, 156)

> Chapter 43 says, "The teaching without words, the advantage of living in nonaction—only few in the world realize it." Here the phrases "teaching without words" and "nonaction" are parallel and indicate a similar attitude. No words is the opposite of many words, just as nonaction is the opposite of assertive action, which also means that many words are an indication of working with assertive action. . . . When chapter 5 says, "Many words lead to quick exhaustion—much better to guard it within," it speaks of the ruler: the more decrees he issues, the faster his reign becomes inflexible. It is much better if he maintains Dao in clarity and stillness within. (1988, 36)

Chapter 51 has, "The veneration of Dao and valuing of virtue is not due to some outside command"—the Mawangdui version using the word "decree" instead. There are two major readings of this line. One says that Dao and virtue are venerated by the myriad beings spontaneously: there is no one agent to hand down an order or arrange for them to do so, but they have done so since time immemorial (Ren 2006, 112; Xu 1985, 25). The other says that Dao and virtue are honored because they do not manage or coerce the myriad beings, but allow them to follow their self-being (Jiang 1988, 316; Chen 2007, 262). The two readings are significantly different, but the second to me seems more appropriate, because all the myriad beings do so without exception. Dao and virtue can only receive veneration because they never order the myriad beings about but constantly let them function in their self-being. The Guodian manuscript has,

> Dao is constant and has no name,
> Simple yet subtle.
> None in heaven and earth dares to make it their subject.
> If the lord or king guards it,
> The myriad beings are self-submitting.
> Heaven and earth rest in harmony
> And send down sweet dew.
> The people, with no one ordering them, are self-balancing. (ch. 32)

This passage makes the close connection and mutual response of Dao with heaven and earth, and ruler with myriad beings and hundred clans amply clear. The last line quite obviously indicates that the people follow their self-being without any outside command or order—and by extension all the myriad beings of whom people form a part. As Jiang Xichang puts it, "This says that the reason why Dao is being venerated and virtue valued lies in the fact that they do not issue orders or coerce the myriad beings, but let them grow and develop of their own accord" (1988, 316). The first reading does not take this close relationship between Dao and beings into account, while the second one does. It clearly points to the interpretation of the passage, "Dao models itself on self-being," as referring to the self-being of the myriad beings—the statement not explicitly using the words "myriad beings," but implying them.

The term "self-being" is not as old as the word Dao, but was first used and developed by Laozi. When he speaks of self-being, he does in no way imply the modern usage of individual or environmental "nature," but entirely focuses on the spontaneous unfolding, the inherent so being of each individual entity. In this sense, the term has a long history in Chinese philosophy, representing the strong imperative that one should not force or coerce but let things be as they are of themselves, in their self-being.

Self-Being as Process

Laozi uses a series of other compounds involving the word "self," such as self-enriching, self-transforming, self-rectifying, self-simplifying, self-balancing, self-submitting, self-generating, self-arriving, and more.[7] The terms are all reflexive in nature and imply both a sense of personal self and of natural process, emphasizing the spontaneous development of beings as well as their personal autonomy and self-determination.

Laozi uses these terms to clarify the lifestyle and attitude of living beings. They do not particularly deal with the nature of Dao or the sage. The most telling example appears in the Guodian manuscript:

> Therefore, the sage says,
> I remain free from affairs
> And the hundred clans are self-enriching.
> I give up all assertive action
> And the hundred clans are self-transforming.
> I cherish stillness
> And the hundred clans are self-rectifying.
> I desire only not to have any desires
> And the hundred clans are self-simplifying. (ch. 57)[8]

Here the sage speaks in the first person, counterpoised to the people. This is also formulated in chapter 32, where "the lord or king guards Dao," and the myriad beings are "self-submitting" and the people "self-balancing." Again, expressions using the term "self" refer to the myriad beings and hundred clans. As the *Zhuangzi* has it,

> Therefore, it is said, those who shepherded the world in ancient times were without desire, the world was satisfied, without action, and the myriad beings were transformed. They were deep and silent and the hundred clans were settled. (ch. 12)

When the text speaks of them being transformed and settled, it means they were "self-transforming" and "self-settling," again using the "self" expression in connection with the myriad beings and hundred clans.

The *Guanzi* similarly notes that as "the ruler remains free from affairs the people are self-examining." It says,

---

[7] Others include also self-awareness and self-love, whose meaning is more positive, as well as self-justifying, self-valuing, self-envisioning, self-attacking, self-preparing, which are seen more negatively.

[8] The transmitted edition reads, "Therefore, the sage says, I rest in nonaction, and the people are self-transforming. I relish stillness, and the people are self-rectifying. I remain free from affairs, and the people are self-enriching; I remain free from desires, and the people are self-simplifying."

> The enlightened ruler, in governing the realm, calms his people and does not trouble them, gives them leisure and does not overwork them. Since he does not trouble them, the people will follow of themselves [are self-following]. Since he does not overwork them, they people will make an effort of themselves. ("Xingshi" 形势; Rickett 1985, 67).

Again, nonaction is the attribute of the ruler above, while "self-following" and "self-examining" describe the people below. The *Shida jing* echoes this, "Forms constantly determine themselves, and for this reason I am more at rest. Affairs constantly manage themselves, and for this reason I rest in nonaction" (ch. 15). Self-determining and self-managing describe the bodies and affairs of the masses, contrasted with ruler's way of life in peace and stillness. The *Huainanzi* says,

> Therefore, the rule of the perfect man is like this: His mind is coextensive with his spirit; his physical form is in tune with his inherent nature. In stillness, he embodies virtue; I movement, he patterns himself on pervasion. He follows his self-being and aligns himself with the inevitable transformations. As he profoundly rests in nonaction, the world is self-harmonizing; as he is tranquill and free from desires, the people are self-simplifying. (8.5; Major et al. 2010, 274) [9]

"Self" and "self-simplifying" here are attributes of the world and the people, while "nonaction" is a characteristic of the ruler.

Based on all this, it is safe to say that in the *Laozi* and other early Daoist texts, the construction "self-X," and thus also the term "self-being," expresses the particular way of being, attitude, and behavior of the myriad beings, the hundred clans, or the people in general. It sufficiently proves that the construction does not apply to Dao, and that Dao is not self-being in itself.

When Laozi speaks of self-being, he never relates it to Dao, sage, or ruler, but always to the hundred clans—a tendency that can also be demonstrated in numerous examples from other sources. Here I will present the *Zhuangzi* and *Wenzi* in some more detail.

---

[9] The *Guanzi* has, "Do not trouble or disturb, but let them be in harmony so they can be self-completing" ("Neiye"). The *Zhuangzi* says, "Let there be no seeing, no hearing; enfold the spirit in stillness, and the body will be self-rectifying. . . . Take care to guard it within yourself, and all beings will be self-strengthening" (ch. 11).

Zhuangzi on Self-Being

The *Zhuangzi* uses "self-being" in two distinct ways. The first is very close to the *Laozi*, as an attribute of the myriad beings that are as they are of themselves and move forward of their own accord. For example,

> The men of old dwelt in the midst of crudity and chaos; side by side with the rest of the world, they attained simplicity and silence there. At that time, yin and yang were harmonious and still, ghosts and spirits worked no mischief, the four seasons kept to their proper order, the myriad beings knew no injury, and living creatures were free from premature death. Although human beings had knowledge, they did not use it. This was the state of perfect oneness. At this time, no one did any assertive actions; all rested constantly in self-being. (ch. 15)

Here, as in other passages cited earlier, self-being refers to a state of the myriad beings, the host of living creatures, as well as the lifestyle of men of old. In another passage, the Yellow Emperor explains how perfect music works.

> Perfect music must first respond to the needs of humanity, accord with the principle of heaven, proceed with the five virtues, and blend with self-being. Only then can it bring order to the four seasons and bestow complete harmony upon the myriad beings. Then the four seasons will rise one after the other, the ten thousand things will take their turn at living. (ch. 14; see Watson 1968, 156)

Here, too, self-being is related to the perception and activities of the myriad beings, the way in which they live naturally as related to perfect music.

> Let your mind wander in mellowness, blend your spirit with vastness, follow along with the self-being of beings, and make no room for personal views—then the world will be governed. (ch. 7; see Watson 1968, 94)

The person addressed in this passage is the ruler, the counterpart of the myriad beings. He is encouraged to rest in stillness, so that all beings and nature can follow their self-being. The same is also expressed in the following:

> The murmuring of water happens in nonaction and self-being. The perfect man stands in the same relationship to virtue: without cultivating it, he possesses it to such an extent that beings cannot draw away from him. It is like heaven is of itself high, earth is of itself thick, sun and moon are of themselves bright. What is there to be cultivated?" (ch. 21; see Watson 1968,226)

The second way Zhuangzi uses "self-being" is in reference to the unadorned, plain, and original nature of beings, their inherent, inborn qualities. In this sense, the term is close to the concepts of heaven and virtue. For example,

> Rites are something created by ordinary men of the world; perfection comes from heaven. Resting in self-being, it cannot be changed. Therefore, the sage models himself on heaven, prizes perfection, and does let allow himself be cramped by the ordinary. (ch. 31; see Watson 1968, 350)

Perfection here is the original aspect of heaven within people, the natural self-being of existence. Another passage speaks more clearly about self-being as self-being, in contrast to evaluations and particular feelings and emotions.

> Huizi asked Zhuangzi, "Can a man really be without feelings?"
> "Of course."
> "But," Huizi asked again, "a man who has no feelings-how can you call him a man?"
> "Dao gave him a face; heaven gave him a form—why can't you call him a man?"
> "But if you've already called him a man, how can he be without feelings?"
> "That's not what I mean by feelings. When I talk about having no feelings, I mean that a man does not allow likes or dislikes to get in and do him harm. He just constantly goes along with self-being and does not try to help life along." (ch. 5; see Watson 1968, 75-76)

Going along with self-being, therefore, means living in alignment with one's heaven-given inherent nature.

In expansion of the *Zhuangzi*, the *Wenzi* speaks of self-being not only as an aspect of the myriad beings, but also presents its application within the affairs of the hundred clans. For example,

> To govern the world according to Dao does not mean to change inherent nature. It means smoothly going along with what it is. Thus, compliance is great; transformation is minute. Those ancient rulers who guided the rivers went along with the natural flow of water. Those who grew plants went along with the demands and circumstances of earth. Those who waged war went along with the will of people.
>
> If one can follow and go along with things, one will have no enemies in the world. As much as all things have their self-being [what is natural to them], all human affairs can be in good order. Thus, the sage makes laws to guide people in their self-being. Both the living and the dead are equally satisfied. (ch. 8, "Ziran" 自然)

> Thus, there is nothing to do about the world except going along with its self-being and enriching and expanding it. There is nothing to do about the change of all things but to follow their essence. So, the sage cultivates himself according to the roots of Dao, never modifying its expressions. He steels his spirit and hides his intellect. He never acts, but everything is achieved; he never governs, but the myriad are in order. What we call nonaction means not to act before the myriad beings act of themselves; what we call nongoverning means not to change the self-being of the myriad beings. The reason that they are in good order is that the sage goes along with their self-being. (ch. 1)

Beyond this, the text also says, "Rectify your way and all beings will rest in self-being" (ch. 2), "Therefore, the sage never acts in pursuit of fame, never uses knowledge in pursuit of reputation, but in governing always follows their self-being, and there is nothing he cannot achieve" (ch. 4). Again, self-being here refers to the natural way of being of the hundred clans and all living beings.

All this further shows that the passage, "Dao models itself on self-being," really indicates that Dao honors and goes along with the natural self-expression of the myriad beings.[10]

Nonaction

The self-being of the passage, "Dao models itself on self-being," therefore does not indicate a particular characteristic of Dao. This is made clear from its juxtaposition with nonaction, in the *Laozi* the key mode of life-style of both Dao and the sage king. The counterpart of their nonaction among the myriad beings and hundred clans is self-being (Ikeda 2009, 527-98). The concept of nonaction in the *Laozi* goes beyond the common usage of self-being in early Chinese literature. Nonaction already occurs in the *Shijing*, as for example, in the following ode:

---

[10] Further instances also appear in the *Huainanzi*. For example, "Each produces what it urgently needs in order to adapt to aridity or dampness. Each accords with where it lives in order to protect against cold and heat. All beings attain what is suitable to them; beings accord with their niches. From this viewpoint, the myriad beings definitely accord with what is natural to them [their self-being], so why should sages interfere with this?" (1.7; Major et al. 2010, 56).

"Heaven extends to the highest; Earth extends to the thickest. The moon illuminates the nights; the sun illuminates the days. Yin and yang transform; clusters of stars are bright. There is no purposeful activity in this. If you rectify their ways, all will be beings in [a state of] self-being" (20.6; Major et al. 2010, 799).

The hare is slow and cautious;
The pheasant plumps into the net.
In the early part of my life, I still rested in nonaction.
In the later part of my life, I met with all these evils.
I wish I might sleep and never wake more. ("Wangfeng")

Here, however, it is a common verb, unlike in the *Laozi* where it is transformed into a highly specialized technical term.

In the bamboo manuscripts, nonaction is mostly but not always written *wangwei* 亡为, literally, "without action." In this form it occurs seven times, each clearly describing a quality of Dao. For example,

Dao is constantly without action.
If lords or kings can guard this,
The myriad beings will be self-transforming,
Transforming yet wishing to do things,
They still maintain nameless simplicity
And always know when it is enough.
Knowing when it is enough to keep still,
The myriad beings are self-determining. (ch. 37)

According to this, Dao is nonaction from beginning to end. The text clearly connects nonaction to Dao, as is also evident in its placement in the "Dao" section of the text in the Mawangdui version. It has some variation, though, emphasizing the quality of namelessness a great deal more. It begins by saying that "Dao is constantly nameless," then, after a few gaps, adds to the line on "nameless simplicity,"

Resting in nameless simplicity,
They will feel no shame.
Free from shame, they can be still.
Then heaven and earth are self-rectifying. (Version A)

Version B matches the Guodian reading, but adds the word "I" after "wishing to do things": "I would describe that as maintaining nameless simplicity. Maintaining nameless simplicity, they will feel no shame." The remainder matches Version A.

The transmitted edition has further variants, omitting the parts on "shame" and adding to the first line, "Dao is constantly in nonaction, and there is nothing that is not done," using "nonaction" for "without action," replacing *heng* with *chang* for "constantly," and generally echoing the Guodian version more while leaving out the notion of namelessness from Mawangdui. Still, even the Guodian version speaks of namelessness, again using *wang* to read literally, "without names."

> Dao is constantly without names,
> Simple yet subtle,
> None in heaven and earth dare make it their servant.
> If lords or kings can guard it,
> The myriad beings will be self-relying.[11]

This shows that Dao is different from concrete, visible, and named things and phenomena, transcendent and imperceptible: Dao is formless, imageless, nameless, simple, and subtle. All these qualities, including nameless simplicity serve to emphasize the wordlessness and purity of Dao—they match its characteristic of nonaction. Saying that Dao is nameless strengthens the contention that it is indivisible and forever unified; saying that Dao rests in nonaction intensifies its particular mode of being.

Another passage that goes in the same direction contrasts "acting for learning" with "acting for Dao," the latter matching the idea of nonaction and implying the continuous diminishing of effort.

> He who acts for learning increases daily.
> He who acts for Dao diminishes daily.
> Diminishing and again diminishing,
> He reaches being without action [nonaction].
> Without action, there is nothing that is not done. (ch. 48)

Scholars used to think that the last bit on nothing not being done was added by later editors, but it already shows up in the Guodian manuscripts and is thus part of the original. The text insists that, in pursuing a life of Dao, one will reach a point where one can be without action and rest in nonaction, which in turn allows things to unfold of themselves so that there is nothing that is not done. This not only shows that nonaction is consistently described as a characteristic of Dao but also that Dao constantly rests in nonaction and thus continues to allow the universe to function. In Laozi's thought, the metaphysical concept of Dao stands in contrast to the concrete entities of the world, the myriad beings. The reason why these beings can unfold in perfect accordance with their inherent nature—self-being, self-transforming—is precisely that Dao remains in its fundamental condition of nonaction.

---

[11] The transmitted edition has, "Dao hides in namelessness" (ch. 41), but this does not appear in the manuscripts.

Huang-Lao thinkers, too, see nonaction as a fundamental quality of Dao and use the term strictly in this sense. For example, the *Daoyuan* has,

> Highest Dao
> Is so lofty it cannot be examined,
> So deep it cannot be fathomed,
> So clear and bright, no one can name it,
> So vast and large, no one can give it form.
> It stands alone and has no opposite,
> None among the myriad beings can order it about. (see Yates 1997, 175)

Nonaction here is on par with other Dao qualities, such as oneness, emptiness, and harmony, used to illustrate its original nature or "plainness." The *Guanzi* makes a similar point, when it says: "Thus, best learn to never speak and maintain nonaction in all affairs. Thereby one can truly know the structure of Dao" ("Xinshu shang"). This emphasizes that people need only to focus on is the principle of non-speech and nonaction to grasp the root core of Dao.

The *Wenzi* similarly calls nonaction the "ancestor of Dao" (ch. 9, "Xiade" 下德), and the *Guanzi* describes it as "a name for Dao itself" ("Xinshu shang"), thus defining Dao particularly through it. The *Zhuangzi* speaks of Dao as having "reality and signs but being without action or form" (ch. 6), which is very close to the *Hanfeizi* which notes that "emptiness, stillness, and nonaction are the reality of Dao" (ch. 8). Both focus on nonaction as the "reality," the essential quality of Dao.

As is also exemplified in numerous Huang-Lao passages, in the metaphysical dimension of Daoist thought, nonaction is the particular characteristic of Dao, defining its mode of existence and function in the world. In the *Zhuangzi*, it also serves as a particular quality of heaven, but this is secondary to its being central to Dao. Here personal freedom and easy wandering are the expression of nonaction within the individual, who "floats about in nonaction as his guide."

Rulership

In human society, moreover, nonaction it is most of all the mark of sagely rulership, the core principle and primary mode of enlightened governing. As Zhuangzi says, "If the ruler finds he has no choice but to direct and look after the world, then his best course is nonaction. Once he has nonaction, he may rest in the reality of his inherent nature and destiny" (ch. 11). In the greater universe, Dao is the single protector and guarantor of the myriad beings; in human society, the sage king is the single protector and guarantor of the people. Confucians and Mohists describe the sage king as the Son of Heaven, which means they see him as the representative of heaven among humanity and thus the chief executive of power.

Laozi, however, does not speak of him as such. In his logic, the sage or ruler, lord or king in his role as master of humanity is intimately connected to, and isomorphic with, Dao as the master of the myriad beings. And since the best possible way of dealing with the myriad beings for Dao is nonaction, so also the best possible way for the sage ruler to deal with the people is to practice nonaction: follow the law and work with Dao. Thus, as the passage cited above, has it, Dao rests constantly in nonaction and lords and kings should match this, thereby to assure that the myriad beings transform of themselves. The transmitted edition uses nonaction only in connection with Dao, sage, and ruler. The Guodian version, too, connects it primarily with the sage king.

> Act on it and you will destroy it;
> Holds on to it and you will lose it.
> For this reason
> The sage is without action and never destroys,
> Is without holding on and never loses. (ch. 29)

> Act without action,
> Deal without affairs.
> Taste without flavors.
> Take big things and make them small.
> Many easy things inevitably grow to be difficult.
> Thus, the sage
> Sees difficulty even in what seems easy,
> And to the end is without difficulties. (ch. 63)

> Therefore,
> The sage rests in affairs without action
> And practices the teaching without words. (ch. 2)

The transmitted edition, too, has several relevant passages:

> Such is the government of the sage:
> He frees their hearts and fills their stomachs.
> He weakens their ambitions and strengthens their marrow.
> He always causes the people to be unknowing and free from desires,
> So that the smart ones will not dare to impose.
> He acts in nonaction, and there is nothing that is governed. (ch. 3)

Loving the country and governing the people,
Can you remain in nonaction? (ch. 10)

The teaching without words,
The advantages arising from nonaction—
Few in the world ever attain them. (ch. 43)[12]

All this makes it clear that nonaction describes the attitude and behavior of the ruler—mostly described as the sage—in the act of governing. In some Laozi passages, moreover, the sage ruler speaks in the first person: "I rest in nonaction, and the people are self-transforming" (ch. 57), or "Because I know nonaction, I find advantages" (ch. 43). Never does the text use nonaction to speak of the people or the hundred clans, nor does it apply self-transforming or self-being to describe the sage or the ruler. The contrast between the two is pervasive and consistent: nonaction is for Dao, sage, and ruler; self-being is for beings and people.

The Overall Structure

The passage, "Dao models itself on self-being," is a lot more complex than it appears at first sight. It cannot be discussed in just one single dimension. Applying the above analysis to the overall structure of Laozi's thought, we can take the discussion one step further and see in more detail that the passage essentially means that Dao goes along with the self-being of the myriad beings.

On the one hand, Laozi's thought deals with the metaphysical understanding of Dao and explores its relationship to the multiplicity of created beings. On the other hand, it discusses questions of political thought and government, centering on the relationship of the sage king and the masses of the people. As noted earlier, scholars cannot imagine that the highest force of Dao would model itself on the self-being of the myriad beings, mainly because Laozi's philosophy presents Dao as the highest metaphysical concept, the creative power that gives birth to the myriad beings. How could this high-level force model itself on or go along with anything else? But this is

---

[12] The *Lunyu* gives an example pertaining to the sage king Shun 舜, revealing some influence from Laozi: "The Master said, 'As for one who has governed in nonaction, isn't Shun a good example? What did he do? He did nothing but gravely and reverently face south on the throne'" (15.5).

precisely where we find the deepest mystery and greatest wisdom of Laozi's philosophy.

There is no question that, in Laozi's thought, metaphysical Dao is the root source of the myriad beings. This means that there can be no division between them in principle. The key passages on this are *Laozi* 42 and 53, the latter stating, "It gives birth to them but does not own them." In addition, the text speaks of Dao as "the mother of the myriad beings" (ch. 25) and as their "most honored one" (ch. 62). From Laozi's viewpoint, Dao not only gives birth to the myriad beings but also serves as the unifying foundation and perpetual guarantor of their continued existence. Thus, for Laozi, Dao contains both, the power of creation and the essential substance of existence.

> Great Dao is all pervasive!
> Found on left and right.
> The myriad beings depend on it:
> It gives birth to them, yet is never stingy.
> It completes its work, yet does not claim possession.
>
> It clothes and nurtures the myriad beings, yet does not lord over them:
> We can call it small.
> The myriad beings return to it, never acknowledging it as their master:
> We can call it great.
> It never, to the very end, thinks of itself as great.
> Thus, it can fulfill its greatness. (ch. 34)

Laozi here clearly says that Dao gives birth to the myriad beings as well as takes care of them and nurtures them in their existence. It plays a huge role in their continued existence: "It is good at imparting what is needed and makes them complete" (ch. 41). In Laozi, Dao is both the original cause of the myriad beings as well as their highest standard and central measure. Understood from this perspective, it is impossible for the myriad beings to go against Dao in their activities and behavior: they are bound to rely on Dao at all times, to go along with it and follow its patterns. "Therefore, relying on Dao in all affairs, one joins Dao . . . Having joined Dao, Dao rejoices in his attainment" (ch. 23); and, "There is none among the myriad beings who does not venerate Dao and value virtue" (ch. 51).

The lines preceding the passage, "Dao models itself on self-being," speak of four great things that successively model themselves on the next: humanity, earth, heaven, and Dao. All these form part of the myriad beings, and they all inherently model themselves on Dao. Just in case that any of them ever exhibits any tendency toward mutation or untoward development, Laozi strongly pushes the notion of "return to the root."

All beings flourish, but each returns to the root.
Return to the root means stillness:
This is called recovering life.
To recover life is called constancy.
To know constancy is called enlightenment.
If you do not know constancy,
You will fall into error and end in disaster. (ch. 16)

On the other hand, and even more importantly, Dao as the mother of the myriad beings also lets each of them unfold and grow freely according to its particular inherent nature, its "virtue." There is thus individual variation, however much the inherent nature of beings is originally bestowed by Dao. Although serving as the great mother to the myriad beings, Dao never claims merit or poses as great; it continues to provide and never insists on pushing its position. From beginning to end, Dao is the most powerful supporter of the myriad beings, playing the role of charitable giver. The myriad beings come in a plethora of different species; they also have vastly different individual natures and continue to develop and manifest in life according to their particular unique characteristics, thereby fully realizing themselves.

This really is the most beautiful virtue of Dao—the fact that it does not exercise any form of manipulation or control over the myriad beings, but lets them develop and unfold as they will. "It gives birth to them but does not own them; it acts on them but does not make them dependent; it grows them but does not master them" (ch. 51). Laozi calls this quality of Dao nonaction. Nonaction does not mean that Dao undertakes no activities or never does anything. Rather, it means that the way Dao acts is free from all manipulation and control. Its sole purpose is to let the myriad beings act in their own way and deal with their affairs as they will. The result of Dao's mode of action, then, is precisely the ongoing emergence and transformation of the myriad beings: "There is nothing that is not done."

The line, "Dao always rests in nonaction," is read by Wang Bi to mean "go along with self-being," more specifically, the self-being of the myriad beings. This does not contradict the idea of Dao as the supporter and protector of all life: since Dao cannot take care of all and everything, the myriad beings have to rely on their own strength to grow and flourish, change and transform. In this sense, the myriad beings are both autonomous and independent, and the nonaction of Dao is to safeguard this! Schematically the relationship can be depicted as follows:

Dao	→	nonaction
↓		↓
myriad beings	→	self-being

Nonaction is the process, the activity mode of Dao; it is how Dao moves in the world. Dao going along with self-being does not refer to its own original presence, but to that of the myriad beings. Self-being in Laozi means that all existing entities grow and develop in accordance with their unique original nature and fully express themselves as who and what they are.

Human society in this context is one particular world created by Dao. The relationship between Dao and the myriad beings in the greater universe, in the limited realm of human governance is echoed by that of the sage ruler and the people. How does Laozi understand this relationship? As described earlier, the sage rests in nonaction while the people follow their self-being. This can be depicted thus:

sage → nonaction
↓ ↓
hundred clans → self-being

The Guodian *Laozi* has the sage say of himself that he remains free from affairs, actions, and desires, while the people enrich, transform, and rectify themselves (ch. 57). This clearly juxtaposes the sage or ruler and the people—the former always being in a state of "non," while the latter are described as "self" doing in some form. The various expressions here closely connect to self-being. In some passages, the sage is also placed in contrast to the myriad beings, such as when they are said to be "self-transforming" if and when the ruler guards Dao (ch. 37) and when the sage is described as able to "teach what cannot be taught" and "support the self-being of the myriad beings without ever needing to act" (ch. 64).[13] While the sage ruler thus governs the world through nonaction, the myriad beings are self-transforming, self-balancing, and self-submitting—all obvious expressions of self-being.

The sage, therefore, models himself on Dao, represents Dao on earth. Human beings as the people form one group among the myriad beings. Their relationship in human society is essentially the same as that between Dao and myriad beings in the greater universe—both reflecting the counter positioning of nonaction and self-being but on different levels and dimensions of reality.

---

[13] This is from version A. Version B has some minor variants, such as using *wu* instead of *wang* for "non," and "study what cannot be studies" instead of "teach what cannot be taught. Having the two versions side by side, thus, provides invaluable insight into the early transmission and editorial process of the text.

The Huang-Lao Expansion

Already around 500 BCE, Laozi issued a vigorous acclamation of non-interventionism, of opposition to all forms of control, of peaceful and stable government. As he says, "Governing a country is like cooking a small fish" (ch. 60); and, "Clarity and stillness create rightness in the world" (ch. 54). He evaluates an entire string of issues—questions and contradictions—connected to human society and governance and finds that they are all because rulers tend to be meddling, controlling, and positioning. Two chapters in particular condemn such behavior.

> The more prohibitive enactments there are in the world,
> The poorer the people will be.
> The more profit-bringing implements there are among the people,
> The more state and clans will fall into disorder.
> The more crafty dexterity there is among humans,
> The more strange contrivances will appear.
> The more flashy display there is in the law,
> The more thieves and robbers will multiply. (ch. 57)

> The people go hungry
> Because their leaders consume huge taxes.
> Thus they go hungry.
> The people are hard to govern
> Because their leaders engage in assertive action.
> Thus they are hard to govern.
> The people make light of dying
> Because their leaders run after the luxuries of life.
> Thus they make light of dying. (ch. 75)

Taking all this into consideration, the core of Laozi's 5000 words is not a political philosophy centering on the ideal of a "small country with few people" (ch. 80). Rather, it is an exposition of the ways by which rulers and social leaders can be least involved and yet most efficient in managing people, society, and resources—no matter what size the state is. This is what Laozi means when he says rulers should "govern through nonaction" while the hundred clans should practice "self-being and self-governing." It is an expansion of his cosmology into the social and political realm—universe and world, cosmos and society being highly isomorphic and structurally parallel.

In the larger scheme of the universe, Dao in nonaction goes along with the self-being of the myriad beings; in human society, the sage-ruler in nonaction goes along with the self-being of the hundred clans. Huang-Lao thought, as it evolved in the latter half of the Warring States period, continues Laozi's vision of the parallel nature of cosmos and society, but with various new additions. A powerful expression of this thinking appears in the *Zhuangzi*,

> Therefore, it is said that the rulers of high antiquity governed through nonaction, merely by channeling heavenly virtue. . . . Those who shepherded the world in ancient times were without desire and the world was satisfied, without action and the myriad beings were transformed. They were deep and silent and the hundred clans were settled. (ch. 12; see Watson 1968, 126)

The *Wenzi*, too, recognizes that the sage's nonaction is rooted in that of Dao. "Thus, the sage applies Dao to manage them, he holds on to oneness in nonaction" (ch. 3). In a later section, Wenzi is in dialogue with King Ping who asks, "How did the kings of old use Dao to govern the world." Wenzi tells him, they did so "by holding on to oneness and resting in nonaction, and thus going along with the transformations of the world" (ch. 4).

Huang-Lao expands Laozi's cosmology and political philosophy in several ways. For one, it strongly emphasizes the idea of Dao as meaning oneness. Dao gives birth to the multiplicity of the myriad beings—but how can this multiplicity remain ordered and unified? To answer this question, Huang-Lao thinkers focus on oneness and duly interpret it as the central "unified body of law," i.e., natural law, the law that is of itself. Oneness pervades the myriad beings and orders them from within; they all live and act underneath it. There is no need for Dao to manipulate or control them: this is its nonaction. In other words, the nonaction of Dao makes it possible for the myriad beings to live naturally, in self-being, under oneness, to grow and flourish in autonomy.

The other expansion in Huang-Lao has to do with the application of a universal law as standard in the realm. In the political oikumene, to maintain unity and order among the people, there must be universal rules and regulations, statutes and guidelines. This is oneness as law among humanity. For Huang-Lao thinkers, the root of this lies in the oneness of Dao as law, in law as part of cosmic self-being. This concept, then, opens a way to make the ideal of the sage as ruling through nonaction concrete in practical measures: the sage as ruler governs through Dao as law.

This new vision of nonaction, unique to Huang-Lao, differs from Laozi, who saw ruling through nonaction mainly as consisting of abstaining from governmental manipulation and control. However, this raises another question: How, without any form of manipulation and control, can one maintain unity and order in society? Laozi never addresses this issue in practical terms, but Huang-Lao thinkers do in their concept of law. Law makes the establishment of social order possible and allows the sage to rest properly in nonaction—now defined as compliance with the law. Because universal law needs particular standards and guidelines so that people can apply it in daily life, all they have to do is obey the rules in all life activities. The enlightened king or sage ruler does not have to do anything special: his main role is to practice nonaction by resting in clarity and stillness.

A third dimension where Huang-Lao expands on Laozi is in the concept of going along, which here comes to mean going along with the people's self-being as expressed concretely in their feelings. "Feelings" indicates the natural emotions of people that gear them to make particular choices, the specific ways in which they pursue benefit and avoid harm to themselves—their "self-acting." The ruler's nonaction, then, is the ability to go along with the people's tendency to "self-act," to work in their own interests. Both the *Huainanzi* and the *Lüshi chunqiu*, texts that reflect Huang-Lao thought to a high degree, present going along as a key philosophical concept. To give one example, the former has,

> When a sage rules the world, he does not change the people's nature but soothes and facilitates the nature that is already present and purifies and enriches it. Thus, following [the nature of things] may be considered great, whereas making [things' may be considered minor.
>
> Yu dredged the Dragon Gate, broke through Yiqu, demarcated the Yangzi, and channeled the Yellow River [so that] they ran eastward into the sea, by following the natural flow of water.
>
> Hou Ji reclaimed the grasslands and introduced tillage, fertilized the soil and planted grain, enabling each of the five grains to grow appropriately, by following the natural propensity of the soil.
>
> Kings Tang and Wu, with 300 armored chariots and 3000 soldiers in armor, quelled the violent and rebellious and brought Xia and Shang under control, by following the natural desires of the people.
>
> Thus, if you can follow [the nature of things], you will be matchless in the world. Now, if things first have what is natural to them [their self-being], afterward human affairs can be governed. (20.10; Major et al. 2010, 803)

Huang-Lao thinkers emphasize that the ruler can fully work with the law and go along with the self-being of the hundred clans because of the supreme will and the impersonal nature of the law (mainly expressed in rewards and punishments) in accordance with human feelings that guide them to pursue benefit and avoid harm. Laws and regulations suffice to let the ruler rest in nonaction and the people remain in self-being. In Huang-Lao political philosophy, then, the particular individual ethics and sagely wisdom as described in

Confucianism have become irrelevant—Aristotle might even say that the law demands that no longer use "wisdom." The *Zhuangzi* makes the same point,

> Therefore, the kings of the world in ancient times, though their knowledge encompassed all heaven and earth, did not lay plans of themselves. Though their power of discrimination embraced the myriad beings, they did not expound any theories of themselves. Though their abilities outshone all within the four seas, they did not act of themselves. heaven does not give birth, yet the myriad beings transform. earth does not sustain, yet the myriad beings are nourished. The emperor and the king do not act, yet the world benefits. (ch. 13; Watson 1968, 144-45)

If, therefore, we say that the Huang-Lao school is essentially "anti-intellectual," then it is mostly against rulers who, rather than working with an objective set of laws, rely on their personal, individual wisdom. This wisdom, according to the school, rather than developing from an intuitive connection with Dao and enhancing universal connection, is limited and should thus be eschewed.

Based on the discussion above, we can see that, however much the Huang-Lao school expands on Laozi's cosmology and political thought, it still essentially revolves around Laozi's most fundamental principles and concepts: universal Dao, sagely nonaction, and the self-being of the people.

Chapter Six

Laozi

"A Great Vessel"

"A great vessel is late to complete" (*daqi wancheng* 大器晚成) is a classic saying and common proverb from *Laozi* 41. Roughly matching the English, "Rome isn't built in a day," its obvious meaning is that it takes a long time to produce a big vessel or, applied to human life, that any great project or personal transformation requires persistence and endurance over a prolonged period.[1] The word used for "late" (*wan* 晚), however, in both the Guodian B and Mawangdui B versions is written differently, using instead "long" (*man* 曼) or "avoid" (*mian* 免). Early editors of the excavated versions replaced these variants with the transmitted "late," Qiu Xigui using "slow" (*man* 慢) (1995, 119). "Late" or "slow" to complete are fairly close in meaning. However, some scholars have come to acknowledge that "avoid" used in the excavated versions cannot be construed in the same way, but essentially means "not" (*wu* 无), rendering the passage, "A great vessel has no completion." If this were the original intention of the text, the common proverb, cited over thousands of years, would in essence be a misreading of the original *Laozi*. What really is the story here?

From "Late" to "No"

Even before the discovery of the manuscripts, Chen Zhu in his commentaries to the *Laozi* and *Hanfeizi* (1934; 1939) expressed doubts about the "great vessel" passage, noting that the word "late" was in fact a loan for "avoid," which in fact means "not." His argument cites the lines before and after:

---

[1] The *Guanzi* uses the expression "great vessel" as a metaphor for the human being, noting that "Guan Zhong is among the wisest men in the world; he is a great vessel!" ("Xiaokuang" 小匡). Confucius disagrees with that assessment, noting that "Guan Zhong's vessel is puny, indeed" (*Lunyu* 3.22).

A great square has no corners.
Great music has faint sound.
A great symbol has no form.
(ch. 41; see Chan 1963, 160)

Closely parallel, all these begin with "great" and have "no" or "little" as the key adverb, suggesting that the next-to-last word in the line about the "great vessel" should also be a negative: "A great vessel has no completion." Han Fei could not realize this, because his "thinking is fundamentally directed toward success and efficiency" (Chen 1934; 1939, 73). Among the silk manuscripts at Mawangdui, version A does not have this chapter at all, except for two characters, but version B uses "avoid"—thus apparently confirming Chen's hypothesis. In his wake, Lou Yulie (1987, 115), Gao Ming (1996, 24-25), Jiang Rui (2000), and several others have all revised the *Laozi* accordingly (e.g., Dong 2000; Hu 2002; Jiang 2000; Liao 2003; Qian 2004). Their arguments follow two lines: phonology and etymology, on the one hand, and structure and meaning on the other.

In terms of phonology and etymology, Jiang Rui (2000) argues that the Mawangdui edition is correct and matches the "long" used in the Guodian bamboo strips, since the two words were homophones in archaic Chinese. The word *man* for "long" had two basic meanings at the time: "far and wide" as well as "not." The *Guangya* 广雅 dictionary of the 3rd century glosses *man* as *wu* 无 ("not") ("Shiyan" 释言), reflecting an earlier reading in the *Xiao Erya* 小尔雅 dictionary of the Han ("Guanggu" 广诂). The word *man* for "late" originally means "slow and lazy" as, for example in the *Shijing* verse, "Shu's horses move slowly" ("Zhengfeng" 郑风), but there is not a single incidence in pre-Qin literature where it matches *man* for "long" and only a questionable case where it might mean "late." It is thus highly unlikely that the *Laozi* would form an exception.

As for the word "avoid" (*mian*) used in the manuscript, Dong Lianchi (2000) points out that phonology suggests a close connection to "not" (*wu*). In high antiquity, "avoid" belonged to the *yuan* 元 rhyme, while "not" was part of the *yu* 鱼 rhyme, both classified as different modes of the *mu* 母 rhyme, which means they were phonetically close. The word "late" (*wan*) does not appear in the literature until the late Warring States, but it shares the same rhyme class as both "avoid" and "long," making it likely that people at the time used the terms interchangeably, an assessment also supported by Liao Mingchun (2003; 2004).

In terms of structure and meaning, Lou Yulie notes that the lines both before and after the passage plus also the verse, "Great governing has no division" (ch. 28), all share the same pattern. They focus on something "great," then characterize it by a negative: square—no corners, music—no sound, symbol—no form. To match this, the line would have to read "vessel—no completion." This means, the Mawangdui version attained it right, using the word "avoid," which is closest to a negative. As regards the term "vessel," a key passage appears in chapter 29: "The world is a sacred vessel." Wang Bi explains this: "Formless and composed, we speak of 'sacred vessel'." In other words, the idea of vessel here is something harmonious and complete—which means that the verse, "A great vessel has no completion" entirely matches the pattern of the neighboring verses (Huang 1995, 63).

Gao Ming (1996) doubts the transmitted edition, noting that the discrepancy of words in the phrase needs further study, but then goes on to accept Lou Yulie's reading, characterizing it as "absolutely correct" and referring to Chen Zhu's interpretation as "accurate." Chen Xiongquan (2000) similarly recognizes that the three words in question were close in archaic Chinese, all indicating a form of negative with regard to the completion of the great vessel.

From the point of view of structure and meaning, this makes sense. "A great square has no corners" means that it is extremely square to the point where the very concept of corners has perished, reflecting the choice of the word "perish" (*wang* 亡) in the bamboo version. "Great music has faint sound" means it is so amazing that it is beyond individual sounds. "A great symbol has no form," means it is meaningful beyond any particular shape and reaches into formlessness. Following this placement of opposition, of being beyond, the reading "A great vessel has no completion," i.e., is beyond ordinary human attributes, is "long" or "far out" (*man*)—the term used in the Guodian manuscript.

In this context, Jiang Rui (2000) notes that Laozi generally emphasizes nonaction and that the term "great vessel" also appears elsewhere:

> The world is a sacred vessel,
> It should not be acted on.
> Act on it, and you will ruin it,
> Hold on to it, and you will lose it. (ch. 29)

This, moreover, means that the world, as a sacred vessel that cannot be managed or run with ordinary human means, should not be classified in opposition to anything, closely matching the four verses in chapter 41. All four, in addition, begin with the word "great." The bamboo *Laozi* has several more. For example,

Great innocence is like blame.
A great square is beyond corners.
Great completion is like lack.
Great fullness is like an open bowl.
Great skill is like clumsiness.
Great expansion is like contraction.
Great straightness is like bent. (see Henricks 2000, 98)

In each case something "great" is expressed through its opposite, which means that reading *man* as "late" or "slow" would make no sense. "A great vessel has no completion," that is, does not mean that some big project is either late or slow to complete, but it indicates that "without completion, there is nothing that is not complete." It is great beyond the limitations of concrete forms and reaches into formlessness.

Liu Xiaogan (2006, 430) follows this reading as outlined by Lou Yulie, Jiang Rui, and Chen Xiongquan and notes the discrepancy of the different editions. Qian Yuzhi (2004) says that if "late" was the operative word in the verse, the neighboring lines should read the same as, for example, in "Great music is late to sound." This, however, makes not sense and does not match Laozi's overall trajectory. He also emphasizes that "non-completion" connects with the philosophy of nonaction, while lateness plays no role in Laozi's thought.

These various views and arguments I like to describe as the position of rejection. That is to say, scholars who support it all have in common that they reject the transmitted version in favor of the manuscripts. However, there is also the position of affirmation, i.e., scholars who accept the transmitted version as accurate. They include Yan Lingfeng (1983), Xu Kangsheng (1985, 11-13), Li Ling (2002, 21), Peng Hao (2001, 93), Guo Yi (2001), Ding Sixin (2010, 319-32), Chen Yongping (2008), and others. What, then, is their line of reasoning?

In Defense of "Late"

Seen from the perspective of the chronological order of the various manuscript and transmitted editions of the *Laozi*, it is not easy to see how the word "avoid" in the relatively late Mawangdui version should be right and the word "late" in all the others be wrong. The Mawangdui version taboos the word *bang* 邦, the personal name of the first Han ruler, Liu Bang (r. 206-195 BCE), but does not taboo *heng* 恒, the name of the fifth ruler, Liu Heng (r. 180-157), which narrows down its date. Before this, the *Hanfeizi* (ch. 21) and the *Lüshi chunqiu* (16.5) both use the word "late." Their testimony strongly suggests that the Mawangdui version does not provide solid grounds for rejecting the transmitted reading. This leaves open the question how the very earliest version we have today, the "long" of the Guodian *Laozi*, fits into all this. Let us begin by looking at the interconnections of the various editions.

Wang Bi's transmitted edition can relate to the Mawangdui version in two ways. One is that he based his reading on it, and the word "late" he uses is a later scribal error, which cannot be confirmed. The other is that he had no access to it and used a different base text, which means that we cannot use Mawangdui to discredit Wang Bi's reliability. The Mawangdui version can relate to the *Hanfeizi* and *Lüshi chunqiu* in two ways also. One is that it is originally based on them, which means that "avoid" is an error for "late," the "sun" 日 radical being left out; the other is that they go back to different base texts, which means both readings are valid.

In the Han dynasty, there were three versions of the *Laozi*: Heshang gong's *Laozi Daode jing zhangju* 老子道德经章句; and Yan Zun's 严遵 *Laozi zhigui* 老子指归—both transmitted in various collections; as well as the *Laozi Xiang'er zhu* 老子想尔注 by the Celestial Master Zhang Lu 张鲁, which has survived in fragments at Dunhuang. They all have "late" when discussing the completion of the great vessel. All this makes it extremely likely that the Mawangdui reading "avoid" is a scribal error. It is really hard to use the single occurrence of a variant to reject such a large number of other instances.

In addition, Han Fei uses concrete examples and stories to illustrate Laozi's abstract theories when speaking about the great vessel and the great sound, he tells the tale of King Zhuang of Chu.

> King Zhuang of Chu ascended the throne and for three years never issued an edict or took any policy measures. One day, the Right Commissioner of the Army, when attending the throne, took him aside and said, "There is a bird perching on a hilltop in the south. For three years, he has not flapped his wings, issued a cry, or taken flight. He sits silently, without utterance. What's his name?"
>
> The king replied, "Not flapping them for three years, his wings and feathers grow long; not flying or singing, he has a chance to observe the people. He may not fly now but once he rises, he soars into the sky; he may not sing now but once he cries, he startles everyone. Leave well alone. I, the king, understand what it means."
>
> Half a year later, the king started to rule. He abolished ten measures and instituted nine new ones; he censored five major vassals and installed six unknown lords: the country was well governed.

> Next, he raised an army and attached Qi, defeating it at Xuzhou. He triumphed over Jin at Heyong and called a conference of the feudal lords in Song, attaining hegemony over the world.
>
> King Zhuang didn't bother about doing minor good, thus he attained great fame. He didn't show his hand early, thus he reached great success. Thus, it is said, "A great vessel is slow to complete; great music is faint to sound." (ch. 21; see Liao 1959, 224-25)

Clearly, Han Fei here intends to show that great success and great fame do not grow overnight but require thorough preparation and consistent effort over a long period of time. He makes the "great vessel" concrete as outstanding success and overarching fame. Showing the "how" of success, he definitely does not speak of "avoid" or "not."

Still, Qian Yuzhi makes the point that the king's actions and Han Fei's interpretation emphasize his "not" doing, his initial avoidance of all actions, which alone makes him strong enough to reach his goals. Qian concludes on this basis that the word "late" is actually wrong. The king in his answer clearly points out that his apparent inactivity serves a higher purpose, saving his strength and preparing him for great things ahead. He needs to let the time ripen and the opportunities align to "soar up" and "startle everyone." Han Fei also emphasizes that the king doesn't bother about minor effects or worry about appearances, thus attaining great success and fame, which closely matches the king's own words.

Wang Xianqian suspects that the world "harm" is a later addition, since leaving it out would make the two sentences parallel, and I think he is right. Han Fei says that the king does not rush to do good in small ways, and because of this can achieve great fame later; he does not show his hand early, and this allows him to reach great success. His lack of rush and visibility provides a concrete example of his behavior during the first three years of his reign, characterized by a massive lack of edicts and policy measures. He acted this manner, not because he was steeped in the philosophy of nonaction, but because he wanted time and conditions to ripen. Wang Xianqian, Liang Qixiong (1985), and Chen Qixian (1974) all interpret the "great vessel" verse this way. Based on this, Han Fei's use of the word "late" is not at all an error for "avoid," but hits the nail on the head.

The *Lüshi chunqiu* also uses "late" and places the verse between two other, parallel lines: "Great wisdom is not formed; great music has faint sound," the first of which does not appear in the *Laozi*. Its chapter title, moreover, is "Pleasure in Success" ("Lecheng" 乐成) and not "Pleasure in Not Succeeding," indicating that it sees the great vessel as being in fact produced.

Its take on the verse is very similar to Han Fei's, and it, too, provides concrete examples of its application.

> It was three generations before boats and carts first appeared that people became accustomed to them. How could the invention of a good thing have been easy? Therefore, rulers who heed their ministers govern nothing well. The establishment of governmental order is caused by the worth of the ruler.
>
> When Wei attacked Zhongshan, General Yue Yang was in command. Having taken Zhongshan, he returned to report to Duke Wen. He noticed the general's look of self-satisfaction in the accomplishment and commanded the office in charge of official documents, saying, "Bring me all the documents submitted by our ministers and retainers."
>
> The official in charge brought him two boxes. Duke Wen ordered his general to look at them. All had criticized the attack on Zhongshan. General Yue turned and withdrew a few steps, then facing north, he bowed twice and said, "The defeat of Zhongshan was due not to my strength but to my lord's merit."
>
> During the attack on Zhongshan, the number of scholars who discussed the matter and warned of its dangers increased daily. If Zhongshan could not be taken, why would two boxes of documents stating the fact be required? A single document one inch long would have convinced Duke Wen that General Yue would lose. Duke Wen was a worthy ruler, but his ministers still disagreed. This is even truer of mediocre rulers. The problem facing them is that they are unable not to act and yet prevent any from dissenting.
>
> As a general principle, when rulers undertake projects that generate no dissent, their humors, goals, senses, and actions do not fail to be right, so which of their subjects would presume to contradict or perversely doubt them? When all are united in an action, there can be no failure. This is how Kings Tang and Wu accomplished their great achievements against the Xia and Shang, how Goujian was able to take revenge on his enemy. If the small and weak can achieve such results through unity in action, how much more can be achieved by the strong and great? (16.5, sect. 4; Knoblock and Riegel 2000, 390-91)

The *Lüshi chunqiu* here documents that, in order to achieve great success or a great victory, one cannot concern oneself with the minor worries of the people, but must be proactive and single-mindedly hold on to one's plans. Eventually achieving great success, one can then relish its pleasure. Chen Qixian notes that this approach and understanding closely match the *Hanfeizi*. He says,

> Ancient thinkers understood sayings like, 'A great vessel is slow to complete' and 'Great music is faint to sound' as meaning that one should not 'bother about doing minor good' and 'should not show one's hand early.' The present chapter [of the *Lüshi chunqiu*] tells stories of Confucius, Zichan, Yue Yang, and more,

all illustrating how people held back and achieved great success—the very same as the tale of King Zhuang of Chu [in the *Hanfeizi*]. (1984, 992)

Both Han-dynasty editions use the word "late." Heshang gong comments, "A person who is a great vessel is like a set of nine jeweled sacrificial tripods—he can never be fully complete." Yan Zun says, "'A great vessel is slow to complete' means that there is nothing that is not present." In addition, Wang Chong 王充 (127-200) in his *Lunheng* 论衡 uses the expression "late to complete," illustrating it with the examples of the speed of growth of different kinds of trees and the quality of their wood.

Maples and cypresses grow fast and their bark never really gets thick and strong. Sandalwoods, on the other hand, take five months to bring forth the first sap and flourish very late in the spring: their wood is so strong and sturdy people use it to make wagon wheels. The Shang had a whole valley planted with mulberries that grew into a great circumference within seven days. They grew fast and were very strong at the same time—that was really strange! A great vessel is late to complete, a previous commodity is hard to attain. Things that do not even last a single day sell fast and easy—like vegetables and fruit. (ch. 40; see Forke 1972)

Along the same vein, the *Hou Hanshu* 后汉书 biography of Liang Yi 郎顗 and Xiang Kai 襄楷 explains the term by noting that to achieve great success one must work hard over months and even years.

I saw senior official Huang Qiong from Jiang Xia, who devoted himself to Dao and relished its arts, lived his life following nature, wore coarse clothes, and cherished treasures. He deeply studied the classics, yet also was full of the courage to participate in politics. The court treated him as a distinguished guest for a time, but later he resigned because of his health.

Laozi said, "Great music has faint sound, a great vessel is late to complete." It takes three years for a qualified man to rule the country. Now all the people of the world think that Huang is good for the government and wonder why we do not call him back. For this reason, it would be best if your majesty called him back to the capital with honor. [2]

To sum up, from the late Warring States through the Han and into later citations of Han works, all agree in the usage of "late." Based on this evidence, the word "avoid" in the Mawangdui manuscript is a scribal alteration,

---

[2] The *Hou Hanshu* edition used here is with commentary by Liu Zhao 刘昭 (Beijing: Zhonghua, 1987).

something that occurs with some regularity as, for example, using "like" (*ruo* 若) for "respond" (*ruo* 诺) or "only" (*wei* 唯) for "although" (*sui* 雖). It does not warrant rewriting the entire tradition.

The Guodian "Long"

However, there is one more question to consider. Should the Guodian variant "long" (*man* 曼) be read "late" (*wan* 晚) or "avoid" (*mian* 免)? All three have highly similar sounds and belong to the same overarching rhyme group. Most scholars think it should be read "late," with Qiu Xigui (1995) opting for "slow" (*man* 慢) and Li Ling (2002, 21) and Peng Hao (2001, 93) following him. Hu Fenna also is in favor of "late," adding that the word occurs only twice before the Qin (2002). As "long" and "avoid" are homophones in archaic Chinese, it is entirely possible that they all in fact mean "late."

The issue gets complex, though, when we look at the deeper implications of the meaning of the three words. Dong Lianchi notes that the word "late" only starts to appear in texts of the late Warring States, citing all kinds of sources that, however, are not of that period with the exception of the *Xunzi* and *Hanfeizi*. Neither the *Mozi* nor the *Zhuangzi* are late Warring States works, nor is the *Zhanguoce* 战国策.

The latter has three instances where it uses the word "late." In its section on the state of Chu, it records a dialogue between Zhuang Xin 庄辛 and King Xiang of Chu. Zhuang Xin says, "I have heard a saying from the state of Zheng: 'See a rabbit and ask for a dog—this is not too late. Lose a sheep and mend the corral—this is not too slow'." The words "late" and "slow" are close in meaning. King Xiang of Chu ruled from 299 to 265 BCE, that is, the middle Warring States.

In its section on the state of Qin, the text cites Cai Ze 蔡泽 speaking to Duke Ying, saying, "Oy! You're seeing this rather late!" This occurred under the rule of King Zhao of Qin (r. 305-251 llBCE), when Cai Ze replaced Duke Ying as prime minister. In its section on the state of Qi, finally, the *Zhanguoce* tells of Han seeking the military support of the King of Qi. Duke Tian summons the prime minister and asks, "Should we support him sooner or later?" Zhang Gai 张丐 answers, "The later you support Han, the more likely it is that it will be weakened and its land invaded by Wei. Much better to get to it sooner." The word for "late" or "later" used here is entirely the same as that in the *Hanfeizi* and *Lüshi chunqiu* but occurs at an earlier date.

The question now is whether the Guodian "long" can be related in any way to the word "slow." "Slow" usually means "tardy" or "laid back," and as a word it does not occur very often in pre-Qin materials. The *Shijing*, as noted earlier, uses it in the verse, "Shu's horses move slowly," which Mao's commentary explains as "relaxed," and indeed the character originally means "lazy" or "at ease," reflecting a common usage at the time. Based on all this, it is not likely that "long" can be read "slow."

This, however, does not mean that "long" actually means "not." "Long" in pre-Qin texts means "extended" or "distant," sometimes also "far out" in the sense of "ethereal," but never "not." The *Guangya* dictionary glosses it in this sense, but it is a later source. None of the scholars who maintain that "long" really means "not" have been able to find even one supporting pre-Qin example. Jiang Rui offers one from Yang Xiong's 扬雄 *Fayan* 法言, a Han text that is close to the Qin, but still not early enough. Nor are there any convincing examples of pre-Qin passages where "long" means "slow"—all those come from the Han or later. Given this evidence, the Guodian word cannot be read "no" and is therefore not a suitable candidate as a precursor for the Mawangdui "avoid." Nor can it be reasonably read "slow." However, it is possibly to understand it as a variant of "late," lending additional support to the transmitted *Laozi* and its citations in the *Hanfeizi* and *Lüshi chunqiu*.

Parallel Passages

The main reason why scholars read "late" as "not" in the verse on the great vessel is its overall meaning in the greater context of the *Laozi*. More specifically, they present three arguments: 1) The word "late" does not match neighboring verses; 2) there are other, similar passages in the text, that require "not;" 3) Laozi's overall philosophy of nonaction suggests the meaning of non-completion in this phrase. But are these arguments really sufficient?

To begin, the verse on the great vessel is framed by three other, similar sentences, which all define something "great" through its opposite. Two of them in fact use "not" (*wu*), defining by negation: "A great square has no corners; a great symbol has no form." However, the third, "Great music has faint sound," does not use full negation, but rather diminution. It is, therefore, not in the same category as the others.

The word used for "faint" (*xi* 稀) basically means "slight" or "quiet," which is not the same as "no sound" at all. The implication of the phrase is that majestic and imposing music is not easily produced. Applied to rulership, this means that overarching government measures are not easily issued. Both come about slowly and increase gradually. Speaking of Dao, the *Laozi* says, "Listen to it and do not hear it: call it faint" (ch. 14), using the same word. It also says, "Faint words: nature's way" (ch. 23). In all these cases, the idea is subtlety and softness, not the complete absence of communication or the other extreme, "Much talk is easily exhausted" (ch. 5). Laozi calls his philosophy the "teaching without words" (ch. 2), which again is not the same as "faint words."

Faint sounds and faint words are one category, different from the pattern of complete negation. It is, therefore, wrong to say that "faint sounds" are in fact, or should be read as, "no sounds." The verse, although in close proximity to the lines about the "great square" and the "great symbol," thus belongs to a completely different category. Similarly, late or slow completion is not the same thing as no completion, and the line about the great vessel, while closely matching the verse on great music, does not belong in the same class as the others.

The remainder of the text, too, supports this understanding. There is only one other verse that has "no" in the same pattern: "Great governing has no division" (ch. 28). The following line, "A great symbol has no form," reads "Dao, hidden, has no name" (ch. 41) but, despite its use of "no" and its parellel four-character structure, it lacks both the word "great" and the inherent definition through contrast. The latter appears variously in the *Laozi*, but does not use the same pattern. For example,

> Therefore,
> The sage places himself in the background,
> But finds himself in the foreground;
> He puts himself outside,
> Yet he always remains.
> Is this not because he has no personal interests?
> For this reason,
> He can complete his personal interests. (ch. 7)
>
> To contract something, first expand it;
> To weaken something, first strengthen it;
> To destroy something, first let it flourish;
> To get hold something, first give it away. (ch. 36)
>
> If the army is strong, it will perish;
> If wood is strong, it will be cut down;
> The strong and the great are inferior,
> The weak and the soft are superior. (ch. 76)

These and other passages in the *Laozi* provide examples of big and great things that are not complete but emerge from negation in a way quite similar to the line about the great vessel. The same chapter, moreover, places it into context:

> Great virtue seems like a valley;
> Vast virtue seems insufficient;
> Steadfast virtue seems unstable;
> True substance seems flawed.
> Great purity seems to be soiled. (ch. 41)

All these share the word "seems" (*ruo* 若) and thus belong to a particular category of statement, also apparent in the following:

> Great completion seems to be incomplete,
> But its utility is unimpaired.
> Great fullness seems to be empty,
> But its usefulness in inexhaustible.
> Great skill seems to be clumsy;
> Great eloquence seems to stutter;
> Great straightness seems to be crooked. (ch. 45)

These lines are significantly different from the statements in chapter 41. While they use opposites in their description of the item in question, they do not emphasize greatness nor do they use complete negation, working rather with appearance and similarity. Thus, we cannot claim that these various examples form part of a single model in the *Laozi*.

The text, moreover, uses the word "great" in different ways. For one, it serves to emphasize burdensome, negative features of life as in, for example, "great trouble" (ch. 13), "great hypocrisy" (ch. 18), "great delusion" (ch. 27), "great hatred" (ch. 79), "great wars" (ch. 30), "great waste" (ch. 44), and the like. For another, "great" appears as a modifier for positive, life-enhancing entities, such as in "great Dao" (ch. 18), "great man" (ch. 38), "great country" (ch. 60), "great state" (ch. 61), "great undertakings" (ch. 63), "great harmony" (ch. 65), and more. However, in none of these Laozi uses the negative *wu* for definition or explication.

The line, "Great completion seems to be incomplete," furthermore, makes it clear that the line, "A great vessel is late to complete," cannot be read, "A great vessel is never complete." The *Laozi* has not a single mention of a vessel that does not need some form of completion. Since "great vessel" and "great completion" are closely connected, we have reason to propose that both terms essentially convey the same idea. In other words, "A great vessel is late to complete" also means "A great vessel is great to complete." Laozi only states that "great completion seems to be incomplete;" he never says, "great completion has no completion."

The Meaning of "Vessel"

A vessel (*qi* 器) then, is a plate, a type of dish, a kind of pottery, a practical tool, something that gets used in daily life. The *Guoyu* notes, "Regulatory and administrative measures balance out in life; small and large vessels are used according to situation" ("Zhouyu" 周语). Vessels come in different sizes, thus there is a distinction between small and big, light and heavy. The *Guoyu* not only emphasizes their difference in shape but also notes their variety in application. The idea of "vessel," moreover, is extended into the human and political realms, so that both people and the world are different kinds of vessels. People accordingly get to be evaluated for their role in life as either small vessels or great ones; the world, in turn, becomes a "sacred" or "significant vessel" to be treated with care.

The *Yizhuan* says, "What is beyond form is called Dao; what is below form is called vessel." "Vessel" here no longer merely indicates a tool of practical living, but includes any object that has form or shape and in intention is close to the idea of "beings." According to Ikeda Tomohisa, the *Yizhuan* inherits the *Laozi* in its understanding of "vessel" (2006). The latter, moreover, uses the term in three distinct ways.

First, it indicates a concrete tool of daily life, used in a variety of ways. For example, "Clay is molded to form a vessel; its use depends on the space that's not there" (ch. 11); "Great weapons are inauspicious vessels, objects people hate; therefore those who have Dao do not use them" (ch. 31); "The more people use vessels of profit, the more troubled the state will be" (ch. 57); and "Let there be ten and a hundred times as many vessels, but let them not be used" (ch. 80). The line in chapter 41 about the "great vessel" being late to complete matches this kind of concrete understanding.

Second, the *Laozi* uses "vessel" as a metaphor. For instance, "Fish should not be taken out the pool; the state's vessels [means] of profit should not be shown to the common people" (ch. 36); and, "Being kind, one can be courageous; being frugal, one can be generous. Never daring to put oneself ahead in the world, one can be a leading vessel" (ch. 67).

Third, in a more philosophical vein, it uses "vessel" to speak about the emergence and growth of beings and the effects of their unfolding. Here we have, "When the uncarved block is broken up, it becomes a vessel. When the sage uses it, he becomes a leading official. Therefore, great wisdom refrains from cutting [things up]" (ch. 28); and, "Dao gives birth, virtue raises, beings shape, and circumstances complete" (ch. 51), i.e., "vessels complete" in the Mawangdui version.

While "vessel" here signifies a somewhat dimension from the other passages, the *Laozi* never argues against completion or fulfillment in any way. Matching this, chapter 41 concludes, "Thus, only Dao is good at providing in the beginning and good at completing." The Guodian version does not have this line after the word "Dao;" the Mawangdui edition leaves out the word "provide" and reads, "Thus, only Dao is good at initiating and good at completing."

In terms of overall meaning, "A great vessel is slow to complete" thoroughly matches several other passages in the *Laozi* that emphasize how big things come about through the gradual accumulation of small steps. For example,

> Prepare for what is difficult while it is still easy.
> Deal with the great while it is still minute.
> Difficult undertakings in the world always start by working on the easy.
> Great undertakings in the world always start by working on the minute.
> Therefore, the sage never strives for the great,
> And for this reason he can complete the great. (ch. 63)

That is to say, anything large or great is the opposite of something small and minute, a great undertaking is the opposite of a small affair, and to get from one to the other, Laozi insists, one needs to plan, act, and complete. He also says,

> A tree as big as a man's embrace grows from a tiny shoot.
> A tower of nine stories rises from a heap of earth.
> A journey of a thousand miles begins with the first step. (ch. 64)

This, too, indicates that the process of completion of something large or far requires the steady, uninterrupted accumulation of effort; it does not come from not completing. The great things, moreover, that come from the uninterrupted accumulation of effort, just like the great vessel, take a goodly amount of time to be produced and completed. These lines are entirely compatible with the saying about the great vessel. Anything "not complete" or completing by non-completion has no application in practical life but is limited to the formless. While the idea of the great vessel, as noted earlier, is originally and fundamentally concrete and part of ordinary life, Laozi does not stop with that, but also points to great undertakings in the abstract. In that context he acknowledges and affirms that even "a great vessel without form" needs completion. One cannot say, "A great vessel has no completion," however much the text may imply that "a great symbol has no symbol."

Jiang Rui and Qian Yuzhi argue that Laozi's concept of nonaction applies to the verse, "The world is a sacred vessel; it should not be acted on" (ch. 29), which means that a great vessel should not be actively produced or modified. Qian Yuzhi (2004) further claims that Laozi's phrase, "Acting in nonaction, there is nothing that is not done" (ch. 48), essentially means, "By not acting, one can reach a state where there nothing that is not done." Laozi's "acting by not acting, undertaking by not undertaking, tasting by not tasting" fundamentally means "any action emerges from nonaction, the completion of any undertaking emerges from not undertaking, anything having taste emerges from the tasteless."

This, however, is not accurate. Nonaction is the core concept of Laozi's political philosophy. It means to behave in a way that does not go against the rules of life, does not interfere with or overly control things. It has nothing to do with "not doing anything," but expresses Laozi's criticism of all kinds of unsuitable, excessive action. When Laozi says that "the world is a sacred vessel" that "should not be acted on" or "held on tightly," he really speaks about how to govern the world on the basis of nonaction. That is to say, "acting in nonaction" means that "governing a great state is like cooking a small fish" (ch. 60), which leads to a political philosophy based on nonaction. Laozi does not mean to say that the various states in the world need no governing at all or that rulers should not do anything; he only says that rulers should govern with appropriate and reasonable control.

Understood in this manner, the concept on nonaction cannot be construed as the basis for saying, "A great vessel avoids completion" or "has no completion." Laozi never really negates or opposes completion. This is evident from several verses:

> Difficult and easy complete each other . . .
> His work complete, [the sage] does not dwell on it. (ch. 2)

> For this reason, he can complete his personal interests. (ch. 7)

> Work complete, undertakings fulfilled. (ch .17)

> A great being, in chaos yet complete. (ch. 25)

> His work complete, [the sage] does not hold on to it . . . .
> Therefore, he can complete great things. (ch. 34)

> He completes without acting. (ch. 47)

Circumstances complete [all beings]. (ch. 51)

Thereby he can complete the great. (ch. 63)

People often fail just as they are about to complete. (ch. 64)

Therefore, one can complete the vessel first. (ch. 67)

His work complete, [the sage] does not claim credit for it. (ch. 77)

In none of these many passages does Laozi show any aversion to completion, on the contrary, he is much in favor of it.[3] Based on all this, it is safe to conclude that in all the various editions of the *Laozi*, manuscript, transmitted, or cited, the line, "A great vessel is late to complete," cannot possibly be read as "avoids" or "has no completion."

---

[3] One exception is the line in the transmitted edition, also found in the *Huainanzi* (ch. 12), "He simply does not fill himself overflowing, thus he can shield himself against renewed completion" (*Laozi* 15). The Mawangdui version and the *Wenzi* (ch. 3) both have this line, replacing "thus" with "thereby," but it does not appear in Guodian.

Chapter Seven

Han *Laozi*

Variants and New Readings

Yet another excavated manuscript, the Han *Laozi* at Peking University, has stimulated the study of ancient Daoist thought in a variety of ways (see Han 2010; 2011; 2012; Zhu 2011). Some variant characters in particular are of great interest, notably those that make a difference to the overall meaning and philosophical interpretation of the text. They constitute variants in both directions: as precursors to the transmitted edition and as inheritors of earlier manuscript versions. In the following, I examine a select few of these variants and present some new readings they suggest, which in some cases go against the traditionally accepted understanding of Laozi's thought.

Namelessness

The fundamental structure of Laozi's thought is the relationship between Dao and the myriad beings. Within this relationship, Dao forms the root of the myriad beings, their creator as well as the underlying cause of their self-being, the reason why they are what they are. As the former, Dao is at the center of Laozi's cosmogony; as the latter, it presents his cosmology. Only in a few places does Laozi describe the how of cosmic unfolding. Most of his text deals with the basic nature and manifestation of Dao: how it benefits the myriad beings it has created and how they in turn maintain Dao in their lives and return to it.

The Han *Laozi* differs from the other versions, both earlier and later, in a number of passages:

> Dao is constantly nameless → Dao is constantly in nonaction (ch. 37).
> Dao is praiseworthy and nameless → Dao is hidden and nameless (ch. 41).

To begin, in Laozi, nonaction is a central concept of great complexity that for the most part is used in conjunction with the ruler, indicating a fundamental governmental principle to be honored by those in power. Its direct application to Dao appears twice in the transmitted edition: "Dao is constantly in nonaction, and there is nothing that is not done" (ch. 37); and

> He who acts in Dao diminishes daily.
> Diminishing and again diminishing,
> He reaches nonaction,
> In nonaction, there is nothing that is not done. (ch. 48)

In the Mawangdui manuscripts, the first passage reads, "Dao is constantly nameless," which means, as Gao Ming originally suggested, that both the inclusion of the term "nonaction" and the following phrase, "there is nothing that is not done"—here and also in chapter 48—were added later (1996, 423-25). However, this understanding proved moot when the Guodian bamboo slips were discovered, where the passage reads, "Dao is constantly in nonaction." Not only does this prove that "nonaction" was not a later addition, but it also raises the question of whether "nameless" was an editorial insertion—as discussed at some length by Zhang Shunhui, Peng Hao, Ding Sixin, and others (see Ding 2010, 87-88).

The Han *Laozi*, like Guodian, having "nonaction" in this place strengthens the contention that this is the original reading. It allows the understanding that Laozi strongly emphasizes nonaction as a key characteristic of Dao, which in turn serves as the guiding principle for rulers in governing the country. That, moreover, is precisely what the text says :

> Dao is constantly in nonaction.
> If lords or kings can guard this,
> The myriad beings will be self-transforming,

Even if the line, "and there is nothing that is not done," found in the transmitted edition, was added later, it still fundamentally follows the Guodian version. But it is more than likely that it was not added later, since it appears in the Guodian version of chapter 48, which in turn is also fully cited in the *Zhuangzi* (ch. 22).[1]

Still, both constancy and namelessness are major characteristics of Dao in the *Laozi*. All editions state, "Dao is constantly nameless" (ch. 32), those after the Han replacing *heng* with *chang*—which can be read as an adverb or also as an adjective, i. e., as yet another way of describing Dao: "Dao is constant and nameless." This echoes chapter 41 of the transmitted edition, "Dao

---

[1] The phrase "there is nothing that is not done" also appears in *Zhuangzi* 20 and 23, as well as variously in the *Wenzi*.

is hidden and nameless." In both passages, namelessness is clearly an attribute of Dao.[2] The words *heng* and *chang*, moreover, are very close, both indicating long-term persistence and steadiness. Both occur in this sense in early literature as, for example, in the *Shijing,* where King Wen says, "They became subject to Zhou; the heavenly mandate is not constant" ("Daya"). Similarly, the *Hanfeizi* has, "Only that which begins with the creation of heaven and earth and neither dies nor declines until heaven and earth disappear can be said to be constant" (20.25; Liao 1959, 194).

The namelessness of Dao is fundamentally different from the "names that can be named" (ch. 1). Compared to concrete things and affairs that can be called by specific names, it is a form of ineffability; related to Dao as ultimate reality it is a unifying appellation, a "constant name." That Dao is constantly nameless means that it is eternal and unchanging but does not have a name as other things and beings. Chapter 1 says, "the nameless is the beginning of the myriad beings;" chapter 37 adds, "If I had to stabilize it, I would use nameless simplicity."[3] Although this does not refer directly to Dao, the connection is made clearly elsewhere (chs. 32, 41).

In chapter 1, namelessness is linked to the beginning of the myriad beings; in chapter 37, it forms part of an admonition to the ruler to govern by using the nameless simplicity of Dao to give peace to the contentious minds of the people. This shows that in Laozi, the namelessness of Dao also has practical value as simplicity free from the quest for fame and fortune. This virtue also plays a role in Sima Qian's characterization of Daoists as people whose "teaching values hiding oneself and being nameless."

This matches Laozi's statement in the transmitted edition, "Dao is hidden and nameless" (ch. 41). Mawangdui A does not have the phrase, and B replaces hidden with [illegible], an obscure character that editors have read as "praiseworthy" (*bao* 褒 /襃). Either way, it has the same meaning as "Dao is constantly nameless." Some editors have read the character as matching "hidden," contrasting

---

[2] Most scholars and commentators read "constant" as an adverb modifying "nameless," but it is equally possible that they are equally placed adjectives describing the absolute nature of Dao.

[3] The word used for "stabilize" (*zhen* 镇) does not mean pressure or control forcefully, but indicates a gentle way of enhancing peace and security. It is used similarly in the *Chuci,* "But seeing how others have fared, I stabilize myself" ("Chousi" 抽思; Hawkes 1959, 167). Also, the *Lüshi chunqiu* says, "Divide the precious metals in the storehouses and distribute the grain in the granaries in order to stabilize and pacify the multitudes" (7.5; Knoblock and Riegel 2000, 186). In addition, the *Shiji* speaks of "stabilizing the country and managing the hundred clans" ("Original Record of Gaozu")

with the word "great" used in the preceding lines; others have linked it to Yan Zun's version, which has "overflowing" (*cheng* 盛) in its place (see Mawangdui 1976b, 48-49). The Guodian bamboo slips have a gap here, and editors have tended to fill it with the word "beginning " (*shi* 始), so that the line reads, "Dao is nameless from the beginning" (Liu 2005, 33-34; Li 2002, 21-24). This is entirely possible, mainly because there is no matching evidence elsewhere in the text. Still, Ding Sixin takes the overall shape of the gap and suggests it is best filled with the word from Mawangdui (2010, 323-25; Ikeda 2011, 271-76).

Assuming that "beginning" is the operative word, the sentence can again be read in two ways: "Dao is nameless from the beginning" or "Dao is beginning and nameless." The latter would refer back to the statement, "The nameless is the beginning of the myriad beings," also echoed in the *Hanfeizi*, "Dao is the beginning of the myriad beings" (ch. 5). The Han *Laozi* has yet a different term here, i.e., "enormous" (*yin* 殷).

Han Wei notes that this is a homophone of "hidden" (*yin* 隐) and is also close to " praiseworthy," serving to explicate the nature of Dao as "great" and "overflowing" (2012, 125). Chen Jian has some reservations about the Mawangdui character and, on the basis of the Han version, thinks it is a scribal error for "enormous." His argument is based on the Mawangdui edition of the *Yizhuan*, which has, "The master said, 'The function of the Changes is to enhance encompassing virtue of Zhou when there is a certain lack of Dao." Already Zhang Zhenglang points out that the word "certain" (*duan* 段) and "enormous" are close, but there is no incidence where the character shows up as a variant of the Mawangdui word. The "hidden" of the transmitted edition may thus well be a version of "enormous" (Chen 2013, 5-6).

The main meaning of "enormous" is very big. The *Zhuangzi* has, "The minute is the smallest of the small, the enormous is the largest of the large" (ch. 17); and, "Its wings are enormous but they attain it nowhere; its eyes are huge but it can't see where it's going" (ch. 20). Based on this, "hidden" could well be a copying error for "enormous" in the sense of extolling the greatness of Dao. However, there are also instances where "hidden" itself is used in the sense of "great" as, for example, in the *Chuci*, "To hang at my girdle the coiling green dragon; to have hidden at my belt the sinuous rainbow serpent" (Hawkes 1959, 290). Wang Yi 王逸 comments that "hidden" here means "great." Based on this, the "hidden" in the transmitted edition may either be read as "great" or as a variant of "enormous." Laozi says, "I do not know its name. If forced to name it, I would call it 'great'" (ch. 25). The phrase "Dao is enormous" would perfectly match this passage.

Still, the greatness of Dao is in no way like the great and small of ordinary things; rather, it is perfect greatness, a fundamental way of being great. Where the Guodian edition has, "Dao is beginning and nameless," the Han version uses "enormous" (or "hidden") instead of "beginning." Both words are adjectives, emphasizing qualities of Dao: as root source of the myriad beings in the former, as the fundamental nature of cosmos in the latter.

Dao Provides

Another variant in the editions is

> Good at beginning and completing [beings] →
> Good at providing for and completing [beings] (ch. 41);

The word used for "provide" (*dai* 贷) appears in all transmitted editions: Heshang gong, Yan Zun, Wang Bi, and Fu Yi, but some versions found at Dunhuang replace it with "beginning." Mawangdui A does not have the line, but B reads, "Good at beginning and good at completing"—used by most editors and scholars to supplement the Mawangdui A and Guodian versions (Liu 2005, 33; Li 2002, 24; Ikeda 2006b, 14-19). Some even apply this to the transmitted edition, amending it to read "begin" for "provide" and "end" for "complete." They claim that "'provide' does not really match the intended meaning, so that commentators old and new need to adjust" (Gao 1996, 26).

The Han version has *shi* 貣 instead of *dai* 贷, which looks very much like it. However, the word is also phonetically very close to *shi* 始 (beginning), with which it shares the same rhyme category (Han 2012, 125). Still, "provide" is not the same as "beginning." Looking at the characters, it is entirely plausible that the Han version *shi* is a copying variant for *dai*. On the other hand, there is no instance in the literature where it actually replaces the word for "beginning." Much more commonly, it actually means "provide." An example appears in the *Qinlü* 秦律, "Among ministers, all magistrates and officials above all must supervise their districts and provide for them ("Canglü" 仓律).

While thus "provide" is not likely to appear as an error for "begin," the opposite may well be true, as Bai Yulan notes (2012, 13). If "provide" is the correct reading, "begin" is certainly vastly different in usage and meaning. "Provide" means to manage and order; "begin" never occurs in this sense. Still, some editors claim, "'Begin' in the transmitted version appears as 'provide.' The two words are phonetically close. As the *Shuowen* says , 'provide' is 'manage' (*shi* 施)" (Mawangdui 1986b, 49).

This contradicts Gao Ming's contention that "'provide' does not really match the intended meaning," but it closely links with Laozi's overall thinking.

According to Gao, "good at beginning and good at completing" really means "good at beginning and good at ending." Does this, then, describe a particular quality of Dao? Or is it what Dao does to the myriad beings? He does not say. Others, however, do. Some, like Xu Kangsheng, Liu Zhao, and Li Cunshan, read it as the former; Ikeda Tomohisa reads it as the latter (2006b, 19). I agree with him. Laozi's Dao obviously means cosmic beginning, being "the mother of the world," but it does not have any sense of ending since it is constant and eternal: "Alone it stands and never changes."

Rather than referring to qualities of Dao, the line is best read as meaning that Dao is good at letting all things and beings begin and end. Isn't the fact that "Dao gives birth to" all things the beginning of all existence? And isn't the fact that it supports the ongoing process of their transformation a form of continued beginning? The text says,

> A journey of a thousand miles begins with the first step. . . .
> The people in their conduct of affairs
> Just about to complete them, constantly ruin them.
> If they were as careful at the end as in the beginning,
> They would not ruin them. (ch. 64)

In this passage, all beginning, completion, and ending are clearly related to human behavior. Dao, on the other hand, is "good at providing for and good at completing [things]," giving expression to a particular attribute of the creative force. This Laozi calls its "mysterious virtue" (ch. 51) or its "clothing and raising of the myriad beings" (ch. 34).

> Dao gives birth to them; virtue nurtures them;
> It grows them and raises them,
> Completes them and matures them,
> Nourishes them and protects them.
> It gives birth to them but does not own them;
> It acts on them but does not make them dependent;
> It grows them but does not master them
> This is called its mysterious virtue. (ch. 51)[4]
>
> Great Dao is all pervasive!
> Found on left and right, the myriad beings depend on it:
> It gives them birth, yet is never stingy.

---

[4] "It skillfully benefits the myriad beings" also appears as, "The Dao of heaven benefits all and does no harm" (ch. 81); and, "Highest skillfullness is like water. The skillfulness of water is that it benefits the myriad beings and does not contend, in occupying, the low place all men dislike. Hence it is close to Dao" (ch. 8).

It completes its work, yet does not claim possession.
It clothes and nurtures the myriad beings, yet does not lord over them:
We can call it small.
The myriad beings return to it, never acknowledging it as their master:
We can call it great.
It never, to the very end, thinks of itself as great.
Thus, it can fulfill its greatness. (ch. 34)[5]

A similar sentiment is also expressed in the *Zhuangzi*:

> Lao Dan said, "The government of the enlightened king? His achievements blanket the world but appear not to be his own doing. He continuously provides for the myriad beings, but the people do not depend on him. With him there is no promotion or fame—he lets all beings find their own enjoyment. He takes his stand on what cannot be fathomed and wanders where there is nothing at all." (ch. 7)[6]

The phrase "continuously provides" in this passage has the same meaning as the word "provide" used in the *Laozi*, indicating the way the enlightened king manages his people. The 7th century Daoist exegete Cheng Xuanying 成玄英, adding to Guo Xiang's 郭象 (d. 312) *Zhuangzi zhu* 庄子注, notes, "Supporting the transformations and enhancing all life, he inspires the people to get rid of all evil. He provides and supplies the myriad beings, bestowing good fortune and goodness upon them." Modern scholars, too, read this to indicate the overall ordering, grace, and beneficial management of the king, applying it to the *Laozi* passage in question. As Chen Guying says, he "manages and transforms the myriad beings equally" (1983, 219). Cao Chuji adds to this by noting that "the enlightened ruler transforms and educates, spreading his grace among the myriad beings" (2000, 113). Fang Yong agrees (2011, 127).

---

[5] Jiang Xichang reads this to mean that "only Dao skillfully benefits the myriad beings and only thorough it can they reach skillful completion of their projects" (1988, 178). Gao Heng interprets, "Only Dao can skillfully manage the myriad beings; it is the only one to skillfully complete them, as well" (1989, 97).

[6] Zhuangzi also has, "The perfect man of ancient times applied Dao as benevolence, rested in righteousness, and wandered in the emptiness of freedom and ease. He ate in the plain and simple fields and strolled in the garden of no bestowal. Free and easy, he was nonaction; plain and simple, he was easily nourished. Never providing, he did not put himself forward. The men of old called him this wandering in perfection" (ch. 14).

And Again

Chapter 1 as found in the transmitted edition is hard to understand. Reading the words literally, commentators throughout history have read it in many different ways. Notably the passage "Mysterious and again mysterious—the gate of all wonders," which in the Han version reads, "Making it mysterious and again making it mysterious—the gate of all wonders," is generally seen to describe Dao, asserting that is deep and amazing, the root source of the wondrous transformations of all things and beings.

Unlike the mainstream, Ideda Tomohisa connects the line to "diminish and again diminish" (ch. 48), "deep and again deep," "spiritual and again spiritual" (*Zhuangzi* 12), and "essential and again essential" (*Zhuangzi* 19), as well as to "think and think again, then think about it some more" (*Guanzi*, "Neiye"). All these share the same sentence structure. I think he is right to see a connection, however he tends to read the core words as verbs—deepen, spiritizing, making essential—to match "making mysterious" as ways of acting for Dao. Like "diminish and again diminish," to him they are ways of pursuing Dao from one's present state and applying a process of reduction, reversal, and negation, until one reaches pure Dao as the ultimate source of all (Ikeda 2006b, 179). This makes a lot of sense, but there are some caveats.

My main issue is the word "it" (*zhi* 之), which may or may not refer to the same thing in all the phrases. In lines that involve obvious action such as diminishing and thinking, "it" looks like a pronoun, indicating an object, but even there it is not clear what exactly it indicates. It may well be a particle to fill a space, with no particular meaning or referent. The Han edition has the longer variant, repeating "it" several times. Cao Feng interprets chapter 1 in connection with chapter 48 and reads: "Making it mysterious and diminishing it," linking the passage to "diminishing it and again diminishing it."[7] He notes—and Ikeda agrees—that the repetition of the particle is no accident and that, since both passages match closely, "mysterious" should be read as a verb, while "it" must be understood as whatever needs to be diminished, negated, or made mysterious to get closer to Dao (Cao 2013).

Ōta Seiken 太田晴轩 interprets the line to mean, "Darken and make it ever more mysterious until it reaches its ultimate," noting the identical pattern in the *Zhuangzi* passages. Cao Feng disagrees with this, reading them less practice-oriented and more metaphysically. The only passage beyond the *Laozi* that might justify using "mysterious" as a verb is the version in the

---

[7] Both Yan Zun and Fu Yi in their versions have the additional *zhi* in chapter 48. The *Zhuangzi* also cites it in this form (ch. 22).

Guanzi, "think and think again." That, to me, is not sufficient to justify this reading. I rather tend to assume that "it" is not a direct pronoun but a filler particle with no concrete object whatsoever. Thus, all these various passages, with the structure, "A and again A," would indicate the intensification of certain attributes: mysterious, deep, spiritual, essential. All these are adjectives, while only "diminish" and "think" are verbs. In Laozi, moreover, the line directly refers to Dao, coming right after, "The Dao that can be told is not the constant Dao; the name that can be named is not the constant name." "Mysterious" is a quality of Dao, its deep and unfathomable nature, its quality as the root source of all beings.[8]

Dao and Image

One further relevant variant in the Han edition is,

> A heavenly image, it has no form → A great image, it has no form" (ch. 41)

Mawangdui A does not have this line, and B reads "heavenly image." One interpretation is that "heavenly" is a scribal error for "great," but since it also shows up in Guodian, as I have explained elsewhere using comparable *Yijing* passages, it may well be the original word. However, at this point, I think a revision is in order, and we must carefully consider Ding Sixin's work (2010, 322-23), who has provided ample evidence that "heavenly" should in fact be read "great," and Han Wei's contention that the Han edition represents an error. Although it seems odd that three ancient manuscripts all use "heavenly," I now believe that the original reading is "great."

Another variant occurs in chapter 62: "Dao is the master [*zhu* 主] of the myriad beings → Dao is venerable [*ao* 奥] among the myriad beings," which

[8] Wang Bi comments: "Names are used to determine forms, and style names are used to designate attributes. To speak of Dao is derived from the fact that absolutely nothing fails to follow it and because, of all the terms that might be used to address the state of no beings and no cause, this one has the broadest meaning. Seeking the reason why this name is assigned to it, we find that it connects to the notion of greatness" (ch. 25; Lynn 1999, 95). "Dao, by being formless and nameless, birthes and completes the myriad beings. They are brought forth and completed in this manner, yet they do not know how it happens. This is mystery. Wonder issues from mystery. Both go back to Dao. As the myriad beings reach completion only after being born in mystery, so they come forth only after being born in nonbeing" (ch. 1; Lynn 1999, 51).

in Mawangdui B reads "inflow" (*zhu* 注), usually considered a scribal error for "master." The editors also adduce the *Liji*, "Therefore people consider him venerable" (ch. 9), to show that the term implies "venerable master" and thus has the same basic meaning as "master." Han Wei reads the Han edition as containing the ancient character 㮡, which shares the same rhyme category as "venerable," and thus concurs with the overall reading.

The question arises, what exactly does it mean that Dao is the master (venerable) of the myriad beings? Chen Guying, following Wang Bi, explains it as someone providing "protection," Xu Kangsheng does not explain the word, Li Cunshan reads it as "ancestor," as does Ikeda Tomohisa, linking it to related passages in the *Laozi*, "Dao is the ancestor [*zong* 宗] of the myriad beings" (ch. 4); the *Guanzi*, "Dao is essential [*yao* 要] to the myriad beings" ("Junchen shang"); and the *Zhuangzi*, "Dao is the ground [*suoyou* 所由] of the myriad beings" (ch. 31).

The world "master" means lord and ruler and in the *Laozi* serves to indicate Dao as the ancestor of all creation. In this sense, the term would refer to the root source or foundation of the myriad beings, matching the related passages in the *Guanzi* and *Zhuangzi*, as well as the *Yijing Xici* line that describes it as, "pivot of emergence, master of growth." On the other hand, the *Laozi* tends to use the word "master" in a different sense, describing Dao as loving and benefiting, clothing and nurturing, protecting and supporting the myriad beings, "but never acting as their master." Here Dao is precisely *not* the master of the myriad beings. Chen Guying reads the original character 㮡 as "start" (*duan* 褍/端), a word phonetically close to both "master" and "venerable" and fundamentally meaning "beginning" (2013, 8-10).[9] All this indicates that the passage essentially means that Dao is the beginning of the myriad beings, closely relating to its quality of namelessness discussed above.

Virtue(s)

The Han *Laozi* also has several relevant variants regarding the understanding of virtue and its application in government. Daoists in general, in contrast to Confucians, tend to illuminate human affairs from a perspective of heaven, focusing strongly on the cosmic dimensions of human life. When they speak

---

[9] "Start" means "head" and "beginning." See *Liji*, "Therefore, human beings are at the heart of heaven and earth, the start of the five phases" (ch. 9). Kong Yida comments, "'Start' is like 'head'." The *Jiyun* 集韵 dictionary defines it as "beginning" ("Hengyun" 桓韵).

of "governing with virtue," they accordingly mean founding one's policies on Dao. This, as much as the Confucian view, is deeply rooted in traditional Chinese thought. As research has shown, even in high antiquity both ruler and leading officials were closely concerned with the concept and pursuit of virtue.

The tradition goes back well into the Three Dynasties, being documented particularly for the Shang and Zhou, as evidenced in many instances recorded in the *Shangshu*. Confucius and his disciples inherited it and cherished the belief that the quality of the country's ruler lay inherently with his personal cultivation of ethics and virtues. In Confucian thought, ethics are never separate from rulership but form the core of official life. All its major ealy thinkers—Confucius, Mencius, and Xunzi—provide guidance for good rulership, focusing strongly on ethical education.

While Daoists tend to be critical of Confucian thought in many respects, they agree that governing has a great deal to do with ethics. Most exegetes agree that Laozi essentially provides self-cultivation guidelines for lords and kings, teaching them the arts of rulership, which centers in many ways on the practice of the virtues. However, how Daoists see the activation of the virtues is quite different from the Confucians. Laozi focuses his ethical guidelines on a higher principle: virtues to him reach into a transcendent realm—the mysterious or constant virtue of Dao. In Daoist eyes, the specific virtues Confucians recommend are the product of a society that has already lost mysterious virtue. Not only are they useless for resolving the issues that plague current social life, but they are themselves the reason why those issues came about in the first place.

This position even appears in the Guodian Laozi, as Ikeda has documented (2011, 473-532). Even though the text does not have the lines, "abandon benevolence, get rid of righteousness," "discard sageliness," and other passages of this kind—found in Mawangdui and the transmitted edition (ch. 19)—it yet clearly sees Confucian virtues as the product of a time after "great Dao was lost." It says,

> Therefore,
> Great Dao was lost,
> And then there were benevolence and righteousness.
> The six relationships were no longer in harmony,
> And then there were filial piety and parental kindness.
> States and clans fell into disorder,
> And then there were loyalty and servitude. (ch. 18)

The word used for "and then" (*an* 安) essentially means "thereupon," "after that."[10] Not actually reading it as *yan* 焉, there is no need to see it always as a question particle—as is documented in chapter 38 of the transmitted edition. However, while the manuscript version does not have this particular passage, it is cited in the *Zhuangzi*:

> So it is said, When Dao was lost, then there was virtue; when virtue was lost, then there was benevolence; when benevolence was lost, then there was righteousness; when righteousness was lost, then there were rites. Rites are the frills of Dao and the forerunners of disorder" (ch. 22; Watson 1968, 235).

This closely matches chapters 18 and 37 in the transmitted edition of the *Laozi* with their understanding that common social virtues only arose after Dao was lost.

In addition, both the Guodian and transmitted versions speak of giving up wisdom, craftiness, and artificiality while discarding understanding, benefits, and conscious planning. They thereby negate and transcend values commonly held in society, especially also by Confucians who favor knowledge and assertive action. This is just one instance among many where Daoists criticize things that Confucians approve. Thus Daoists say,

> Those of high virtue did not care about virtues,
> Thus, they had real virtue.
> Those of low virtue lost virtues,
> Thus, they had no real virtue.
> Those of high virtue rested in nonaction,
> Thus, they did not need to do anything.
> Those of low virtue used assertive action,
> Thus, they always needed to do things. (ch. 38)

The most immediate early reference to this chapter appears in the *Hanfeizi*, which comments on it (ch. 20). It is not merely a matter of extract-

---

[10] The text also says, "Men appoach him and come to no harm, and then they can live in peace" (ch. 35). Again, *an* is the connective particle. Wang Yinzhi also reads it this way. The same usage also appears in the *Zhanguoce*, "Therefore they sat for a long time, and then discussed the complaints of the three states." The *Zhuangzi*, moreover, cites the passage from the *Laozi*, but changes it into a question, "If Dao and virtue had not been lost, how would people ever have adopted benevolence and righteousness?" (ch. 9). Here, *an* is a question particle, meaning "how" or "why." The other usage of *an* in the *Zhuangzi* is more like *er* 而: "Do away with and give up benevolence and righteousness, and then the virtue of the world will finally begin to be mysterious" (ch. 10).

ing certain sentences, even if it is possible that some parts are later interpolations. Still, since the line on "lost virtues" appears in the manuscript as well as the transmitted edition, the entire argument is likely to be old.[11] In addition, there are some variants. The transmitted edition says that "they always needed things to do," while the Guodian version has, "they did not need to do anything." (The Mawangdui manuscripts do not have the line.)

The divergence raises doubts. The text variously speaks of high virtue, high benevolence, high righteousness, and high rites, and only once of "low." This is a major mismatch. Only virtue is divided into high and low, all others lack this distinction. Also, the differentiation between high and low virtue is not clear. Is low virtue less than high benevolence? Is it on the same level as high righteousness? Is lower virtue a value in Laozi? Probably not. He consistently proposes a level of Dao and virtue that is higher than that of Confucianism and criticizes the latter's virtues as the product of a time after Dao is lost. Beyond that, he also provides a variety of attributes to virtue, such as high, vast, and established (ch. 41), mysterious (chs. 10, 51, 65), open (ch. 21), present (ch. 79)—all positive and praiseworthy characteristics which make virtue something that "none of the myriad beings do not value" (ch. 51).

Yet, however high it may be in human society, when compared to Dao itself, virtue occupies a lower position. After all, "when Dao was lost, then there was virtue" (ch. 38). This makes the concept of low virtue rather superfluous, especially since there are practical virtues such as benevolence, righteousness, and rites to fill the gap, as described in the *Zhuangzi* (ch. 22). These various virtues crop up when "great Dao is lost," not when Dao and virtue are both gone.

Desires

In another passage about rulership, the text says, "One who has Dao does not dwell [on them]" (*youdaozhe buchu* 有道者不处; chs. 24, 31), which in both Mawangdui versions reads, "One who has desires does not rest [in them]" (*youyuzhe fuju* 有欲者弗居). Most commentators think the second reading is the more original, adducing various arguments. But is that really true? The main difference is just one word, i.e., "Dao" versus "desires," but their meaning could not be more at odds.

[11]The line is also cited in the *Wenzi* (ch. 6) and in several chapters of the *Shiji*.

The Laozi tends to view desires rather negatively, speaking mostly of being free from desires, having no desires, or desiring to have no desires. It essentially never speaks of having them, and certainly not in the context of resting. The sage should make sure "the people have no knowledge and remain free from desires" (ch. 3); because Dao is said to be "always free from desires, it can be called minute;" the ruler should "desire to have no desires" (ch. 34); and so on. It also says, "there is no greater sin than being full of desires, no greater fault than the desire to attain [things]" (ch. 46). Most clearly, the text has,

> Transforming [things] with the desire to act on them—
> If I had to stabilize this tendency, I would use nameless simplicity.
> Stabilizing it with nameless simplicity
> Means to reach a state of not desiring.
> Not desiring anything, one can rest in stillness,
> And the world will be self-rectifying. (ch. 37)

In all these instances desires are something negative that one should overcome and eliminate. Other passages also emphasize humility, simplicity, aloneness, and other forms of no desires. The text never speaks of many desires, things to be desired, or in any other positive way about them. Laozi consistently values having few or no desires; he never encourages the development of having or increasing them.

He also postulates that is good to reduce the kinds of things that might inspire desires in people and encourages the cultivation of moderation (named as one of his Three Treasures; ch. 67) as well as an overall simplicity of lifestyle. He contends that abundance, luxury, and sensory indulgence diminish people's ability to live in self-being: "The five colors make the eyes go blind . . ." (ch. 12).[12] Moreover, he sets up Dao as the highest standard and

---

[12] Xunzi criticizes this aspect of Laozi's thought: "As a general rule, those who contend that political order requires that we first rid ourselves of desire are those who lack the means to guide their desirs and so are embarrassed by their having desires. As a general rule, those who contend that political order requires that we first reduce the number of desires are those who lack the means to moderate their desires and so are embarrassed that their desires are numerous.

"Having and not having desires belong to different categories, that is, those of life and death, and not those of order and disorder. The quantity of desires, whether we have few or many, also belongs to a different category, that is, the calculation of our inherent nature, and not to that of order and disorder. Desire does not depend on the object of desire first being attainable, but what is sought after follows after what is possible.

value for humanity, and consistently admonishes people to cherish and venerate it. With the exception of one passage, "Who can have abundance and serve the world? Only one who has Dao" (ch. 77), which directly uses the expression "have Dao," he always speaks of "having virtue." That, however, as pointed out in the *Zhuangzi*, has a tendency toward decline:

> The time came, however, when virtue began to dwindle and decline, and then Suiren and Fuxi stepped forward to take charge of the world. As a result, there was compliance but no longer any unity. Virtue continued to dwindle and decline, and then Shennong and the Yellow Emperor stepped forward to take charge of the world. As a result, there was security but no longer any compliance.
>
> Virtue continued to dwindle and decline, and then Yao and Shun stepped forward to take charge of the world. They set about in various fashions to order and transform it, and in doing so defiled purity and shattered simplicity. Dao was pulled apart for the sake of goodness; virtue was imperiled for the sake of conduct. After this, inborn nature was abandoned and minds were set free to roam. (ch. 16; Watson 1968, 172)

In many other passages, Laozi speaks of pursuing or following Dao. Examples include:

> Those of old good at acting in Dao . . .
> Preserved this Dao and never desired abundance. (ch. 15)

> Therefore, those who follow Dao also become one with it. (ch. 23)

> A ruler who acts in Dao to help people
> Never forces the world with arms
> And his affairs have good outcomes.
> Wherever troops are stationed,
> Briars and thorns spring up.
> In the wake of big armies,
> There are sure to be bad years. (ch. 30)

> He who acts in Dao diminishes daily.
> Diminishing and again diminishing,
> He reaches nonaction. (ch. 48)

"That the occurrence of desire does not depend on its object first being attainable is a quality we receive from nature. That what we seek to satisfy our desires by following after what is possible is what we receive from our mind. It is natural to our inborn nature to have desires, and the mind acts to control and moderate them" (22.5a; Knoblock 1994, 135)

If the word is managed with Dao,
No ghosts appear as spirits. (ch. 60)

When the ruler has appointed the three dukes,
Even if offered a jade disk and team of horses,
This is not as good as sitting and receiving Dao.
Why did people of old value Dao like that?
Was it not because pursuers can attain it
While the guilty can avoid it?
Thus the world values it. (ch. 62)

Those of old good at acting in Dao
Did so not to enlighten the people,
But to keep them simple and ignorant. (ch. 65)

These passages all specify positive ways of interacting with Dao: good at acting in it, pursuing it, following it, preserving it, and so on. They closely echo the line, "One who has Dao does not linger," but have no correspondence with its alternative reading, "One who has desires does not rest." The context, moreover, clarifies the issue further.

Who is on tiptoes does not stand well;
Who is on stilts does not walk well.
Who displays himself does not shine;
Who asserts himself is not distinguished.
Who vaunts himself has no merit;
Who takes pride in himself has no superiority.
Compared to Dao, these are like spoiled food or cancers—
All beings hate them.
Therefore, one who has Dao does not dwell [on them]. (ch. 24)

Arms are instruments of bad fortune
All beings hate them.
Therefore, one who has Dao does not dwell [on them]. (ch. 31)

In these two chapters, like in many others of the text, Laozi criticizes forms of evil human behavior and actions that go against Dao. One who has Dao naturally would avoid or transcend these things. In chapter 24, he explicitly uses Dao as a standard for human actions and compares certain behaviors to extraneous and disgusting things like spoiled food or cancers. This fully matches the postulate that only one who has Dao "can have abundance and serve the world" (ch. 77). All this makes it highly unlikely that the original passage was, "One who has desires does not rest in them."

Another dimension of desires appears in chapter 1, which has, "Constantly without desires, see their wonder; constantly having desires, see their manifestation." The chapter has a number of tricky passages, which have to remain unexplored at this point. I would rather focus on the chapter as a whole. The lines in question can be read as translated above or, alternatively, "In constant nonbeing, desire to see their wonder; in constant being, desire to see their manifestation." The dominant tendency, both among medieval commentators and modern scholars, has been to read them the first way, matching the Mawangdui version, which adds the final particle *ye* 也 after "desire." Still, even this may be a later addition.

The key words in the section are have/being and not have/nonbeing, central concepts in Daoist thought overall. The *Zhuangzi* describes the philosophy of Laozi and Guanyinzi as "founded on constancy, being, and nonbeing" (ch. 33), which increases the likelihood that chapter 1 should be read, "in constant being/nonbeing." In addition, the chapter speaks of nonbeing as "the beginning of heaven and earth" and of being as "the mother of the myriad beings."

Also, the phrase "desire to" is extremely common in pre-Qin works. For example, the *Zuozhuan* has, "He desired to attack Qi" (Heng 18), while Mencius says, "The fact is that Confucius desired to go away on occasion of some small offense, not wishing to do so without some apparent cause (6B26). Similarly, the *Zhuangzi* uses it in, "Only in their likes were they different from the true sage; what they liked they desired to make clear" (ch. 2); and, "I desired to teach him and see if I could get anywhere near to making him a sage" (ch. 6). The *Hanfeizi* states, "Scholars who have Dao cherish its arts; they desire to make the clear among rulers of a myriad chariots" (ch. 34). All these and many others match Laozi's usage of the word "desire."

Another common pattern in the *Laozi* is the juxtaposition of a quality or term and its negation or opposite. For example, "The good is the teacher of the not good; the not good is the helper of the good" (ch. 27). Mawangdui B leaves out the negation in the second term, which can either be read as is or amended on the basis of the transmitted edition. The Han version matches Mawangdui B, which causes Han Wei to think that the transmitted edition has it wrong. On the other hand, it is also quite possible that both early manuscripts are missing the word "not." The whole point of the chapter is the contrast between the good and those that are not, so leaving the negation out would render its content moot.

Other chapters, moreover, have the same contrast. "Dao is treasured by the good and maintained by the not good" (ch. 62); "The good, I treat with goodness; the not good, I treat with equal goodness; the honest, I treat with honesty; the dishonest, I treat with equal honesty" (ch. 49). These passages all share a certain abstract logic and content structure, which would be lost without the negation: the good would not be the helpers of the good, or their students. If the good were the teachers of the good, the text should make sure to distinguish them in some particular way, so they are not all the same "good" people. However, to read that good people are the teachers of the not good—now that makes sense of the chapter and matches the remainder of the text.

One point of these passages is to show Laozi's great care and compassion, his heart that embraces all the myriad beings. He does not see or treat those not so good differently from the good, the honest differently from the dishonest. As a result, his virtue is full of goodness, full of honesty, always ready to serve and aid and come to the rescue of all.

> The sage in constant goodness rescues the world,
> Never abandoning any one person.
> In constant goodness, he rescues all beings,
> Never abandoning any one being. (ch. 49)

> Even those among men who are not good
> How could he ever abandon them? (ch. 62)

Sageliness

The highest principle and system of governing rulers are to apply according to the *Laozi* center on Dao, nonaction, and the like. As the text says, "To him who holds on to the great image, the whole world repairs" (ch. 35). Here the "great image" is an entity of form that stands for Dao at the root of all. Holding on to it means that the ruler grasps, maintains, and applies the governing principle of the great image, that is, of Dao. As a result, everyone and everything in the world will return to and follow him.

The word "hold" (*zhi* 执) in the Mawangdui and transmitted editions is written slightly differently in the Guodian version, namely 埶. Qiu Xigui reads this to mean "establish" (*she* 设), and many scholars agree with him. Li Ling considers the word merely an odd version of "hold." The variant also appears in the Han version. Han Wei reads it as "establish" and takes it as a variant of "hold" (2010, 161). Ding Sixin reads even the Guodian version to mean "hold" and adduces ample proof for this position (2010, 387-88).

By the same token, the Han version should be read "hold." I think that since the object of the word is the "great image," which stands for Dao, it is highly unlikely that the sage or ruler "establishes" it. Dao at the foundation of the cosmos is an objective, preexisting reality that cannot be said to be "established" by a person, however sagely. As Chen Xiyong emphasizes, Dao in Laozi is formless and nameless; it can be guarded but not established (2005, 277). This is my point. Another way of looking at the character variant, moreover, besides being an odd version, is, as Guo Xiyong suggests, that it is simply a scribal error.

Wang Bi's transmitted edition has, "The sage is has no constant mind; he makes the mind of the hundred clans his own" (ch. 49). In Gu Huan's version as well as in the texts recovered from Dunhuang and in Chen Long's inscription, the first part reads, "The sage is [in a state of] no-mind." Not found in Mawangdui A, it appears in B as well as in the Han edition as, "The sage is constantly in no-mind." This is close to the common variant of the transmitted version.

The text here expresses Laozi's ideal on how the sage ruler should work with the people. As commonly known, a strong emphasis on the will and mind of the people is characteristic of Confucian thought. Already the *Shangshu* has, "Heaven sees itself as the people see; heaven hears itself as the people hear" ("Great Declarations;" cited in *Mengzi* 5A7); and, "Whatever the people desire, heaven must follow" (ibid., cited in *Zuozhuan*, Xiang 31). This emphasizes the degree to which heaven sees and hears through and with the people and shows just how much people-based thinking is part of traditional Confucian thought.

However, Laozi, too, strongly focuses on the will and mind of the people, as the above passage documents. His contention is that the sage ideally does not have any personal will, desire, or inclination, but from beginning to end reflects those of the hundred clans making them completely his own. Thus, whether he rests in a state of "no-mind" or has "no constant mind" of his own makes very little difference in terms of meaning. Even if he has a certain intention or preference, he will never firmly insist on following it but always go along primarily with the wishes of the hundred clans, transforming his mind along with theirs.

The way of rulership in Laozi, therefore, centers on the Dao of heaven and the will of the people as well as on historical models, honoring and emulating the laws of old. The latter is particularly evident from the text's tendency to attribute governmental wisdom to sages of high antiquity and the general Daoist way of establishing systems based on models of old. As the *Hanshu* says, "Followers of Dao emerge from historical officials. History records how they established Dao in antiquity and today, rising and falling, leading to good and bad fortune" ("Yiwen zhi"). According to this, the Daoist way of rulership goes back far in history. This is echoed in the text,

> Hold on to Dao of old,
> To direct the things of the present.
> Knowing the beginnings of old,
> Can be called being a mainstay of Dao. (ch. 14)

However, the Mawangdui versions both have, "Hold on to Dao of today," giving the passage a different slant. Gao Ming thinks this is the original intention of the text and that Wang Bi's edition has it wrong (1996, 288-89). Liu Xiaogan considers the transmitted edition correct, presenting several related passages as evidence.

> Those of old who were good at acting in Dao. (chs. 15, 65)

> Those of old who could be said to value Dao. (ch. 62)

> The utmost of antiquity. (ch. 68)

All these paint old timers and antiquity in a highly positive light, as something to emulate and follow, and document that Laozi was not inclined to elevate the present and downplay the past (Liu 2006, 187). Ikeda Tomohisa argues from the overall flair of the text that the Mawangdui version cannot be right. I agree with this. As I discussed in an earlier article, Laozi is deeply steeped in history and consistently holds a position of emulating the past (see Wang 2002). The Han edition, which has, "Hold on to Dao of old," is just one more item to substantiate this conclusion.

Conclusion

In the above pages, I have elucidated a number of variants and new readings found in the recently discovered Han version of the *Laozi*, notably regarding its two major dimensions, the understanding of Dao and its application in rulership. As has become clear, whenever looking at these newly discovered versions, we must first clarify whether what appears to be a variant really is one, then examine what impact, if any, it has on the intended meaning of the text. Once we have seen this, we can evaluate how the variant offers a new reading and how that new reading impacts our overall understanding. For the most part, it seems, the variants in the Han version are not really variants, since they also appear in other versions. Plus, for the most part, they do not create a major discrepancy in the text's intention. Still, it is important to examine the new versions with their variants carefully and decide, on a case-by-case basis, whether there are differences or similarities to the established editions. The Han version as held at Peking University, in particular, stimulates a new look at the ancient classic and offers a variety of inventive venues for discussion and appreciation.

Laozi Chapters Cited

Bibliography[1]

Addiss, Stephen, and Stanley Lombardo.* 1993. *Tao Te Ching*. Indianapolis: Hackett Publishing Company.

Asano Yuichi 浅野裕一. 2005. "*Kōsen* no dōka tokushoku" 恒先の道家特色. In *Chikukan ga kataru kodai Chūgoku shisō* 竹简が语る古代中国思想, edited by Asano Yuichi. Tokyo: Kiko shoin.

_____. 2009. "Fanwu liuxing de jiegou xinjie" 凡物流形的 结构新解. www.bsm.org.cn/ show_article.php?id=986.

Bai Xi 白奚. 1998. *Jixia xue yanjiu: Zhongguo gudai de sixiang ziyou yu baijia chengming* 稷下学研究:中国古代的思想自由与百家争鸣. Beijing: Sanlian.

Bai Yulan 白于蓝. 2012. *Zhanguo Qin-Han jianbo gushu tong jiazi kuangji* 战国秦汉简帛古书通假字汇纂. Fuzhou: Fujian renmin.

Bowler, Peter J. 1989. *The Invention of Progress; The Victorians and the Past*. Cambridge, Mass.: B. Blackwell.

_____ 鲍勒. 1999. *Jinhua sixiang shi* 进化思想史. Translated by Tian Luo 田洺. Nanchang: Jiangxi jiaoyu.

Brindley, Erica F., Paul R. Goldin, and Esther S. Klein.* 2013. "A Philosophical Translation of the *Heng Xian*." *Dao: A Journal of Comparative Philosophy* 12.2:145-51.

Cai Shuheng 蔡枢衡. 1983. *Zhongguo xingfa shi* 中国刑法史. Nanning: Guangxi renmin.

Cao Chuji 曹础基. 2000. *Zhuangzi qianzhu* 庄子浅注. Beijing: Zhonghua.

Cao Feng 曹峰. 2006a. "*Hengxian* zhengzhi zhexue yanjiu" 恒先政治哲学 研究. In *Shangbo Chu jian sixiang yanjiu* 上博楚简思想研究, 123-48. Taipei: Taiwan wanquan lou tushu fufen yougen gongsi.

_____. 2006b. "Hengxian de bianlian yu fenzhang" 恒先的编联与 分章. In *Shangbo Chu jian sixiang yanjiu* 上博楚简思想研究. Taipei: Taiwan wanquan lou tushu fufen yougen gongsi.

_____. 2006c. "*Hengxian* shiyi siti" 恒先释义四题. In *Shangbo Chu jian sixiang yanjiu* 上博楚简思想研究. Taipei: Taiwan wanquan lou tushu fufen yougen gongsi.

_____. 2009a. "Fanwu liuxing zhongde zuoyou zhiqing" 凡物流形中的 左右之情. www.jianbo.org/admin3/2008/caofeng008.htm.

[1] Items marked * were used by the translator.

______. 2009b. "Yizhou shu, Zhouzhu jie kan Fanwu liuxing de sixiang jiegou" 从逸周书,周祝解看凡物流形的思想结构. www.jianbo.org/ showarticle.asp?articleid=1615.

______. 2010a. *Chudi chutu wenxian yu xianqin sixiang yanjiu* 楚地出土 文献与先秦思想研究. Taipei: Taiwan shufang.

______. 2010b. "Shangbo Chu jian Fanwu liuxing de wenben jiegou yu sixiang tezheng"上博楚简凡物流形>的文本结构与思想特征. *Qinghua daxue xuebao* 清华大学学报 2010/1.

______. 2013. "Xuan zhi you xuan zhi he Sun zhi you sun zhi" 玄之又玄之和损之又损之. *Zhongguo zhexue shi* 中国哲学史 2013/3.

______. 2014. "*Taiyi shengshui* xiaban bufen shi yige duli wanzheng de pianzhang" 太一生水下半部分是一个独立完整的篇章. *Qinghua daxue xuebao* 清华大学学报 2014/2.

Capra, Fritjof. 1975. *The Tao of Physics: An Exploration of the Parallels between Modern Physics and Eastern Mysticism*. Boulder, Col.: Shambhala.

______ 卡普拉. 1999 *Wulixue zhi dao: Jindai wulixue yu dongfang shenmi zhuyi* 物理学之道:近代物理学与东方神秘主义. Translated by Zhu Runsheng 朱润生. Beijing: Beijing chubanshe.

Chan, Shirley.* 2014. "Oneness: Reading the 'All Things Flow in Form' (*Fanwu liuxing*) of the Shanghai Museum Bamboo Texts." *Journal of Chinese Philosophy* 41. www.academia.edu.

Chan, Wing-tsit.* 1963. *A Source Book in Chinese Philosophy*. Princeton: Princeton University Press.

Chang, Leo S., and Yu Feng.* 1998. *The Four Political Treatises of the Yellow Emperor: Original Mawangdui Texts with Complete English Translation and an Introduction*. Honolulu: University of Hawaii Press.

Chen Guying 陈鼓应. 1983. *Zhuangzi jinzhu jinshi* 庄子今注今译. Beijing: Zhonghua.

______. 1984. *Laozi zhushi ji pingjie* 老子注译及评介. Beijing: Zhonghua.

______. 1992. *Lao-Zhuang xinlun* 老庄新论. Shanghai: Shanghai guji.

______ 2007a. *Huangdi sijing jinzhu jinshi: Mawangdui Hanmu boshu* 黄帝四经今注今译: 马王堆汉墓帛书. Beijing: Shangwu.

______. 2007b. *Laozi jinzhu jinshi* 老子今注今译. Shanghai: Yinwu yinshu guan.

______. 2015. *Rediscovering the Roots of Chinese Thought: Laozi's Philosophy*. Translated by Paul D'Ambrosio. St. Petersburg, Fla.: Three Pines Press.

Chen Jian 陈剑. 2013. "Han jianbo Laozi yiwen lingli" 汉简帛老子异文零札. In *Jianbo Laozi yu daojia sixiang—Guiji xueshu yantao hui lunwen ji* 简帛老子与道家思想" 国际学术研讨会论文集. Beijing, 2013/10.

Chen Jing 陈静. 2004. "*Huainanzi* yuzhou shengcheng lun de lilun qianshi: *Hengxian* jiedu" 淮南子宇宙生成论的理论前史: 恒先解读. In *Ziyou yu shixu de kunhuo: Huainanzi yanjiu* 自由与秩序的困惑:淮南子研究, edited by Chen Jing. Kunming: Yunnan daxue.

Chen Ligui 陈丽桂. 2004. "Shangbo jian, san: *Hengxian* de yili yu jiegou" 上博简,三: 恒先的义理与结构. www.bamboosilk.org/showarticle.asp? articleid=1018.

Chen Qixian 陈奇猷. 1974. *Hanfeizi jishi* 韩非子集释. Shanghai: Shanghai renmin.

_____. 1984. *Lüshi chunqiu jiaoshi* 吕氏春秋校释. Shanghai: Xuelin.

Chen Wei 陈伟.1999. "*Taiyi shengshui* kaoshi" 太一生水考释. *Gu wenzi yu gu wen xian* 古文字与古文献. 1999/10.

_____. 2000. "*Taiyi shengshui* xiaodu binglun yu *Laozi* de guanxi" 太一生水校读并论与老子的关系. *Gu wenzi yanjiu* 古文字研究 22.

_____. 2002. *Guodian zhushu bieshi* 郭店竹书别释. Wuhan: Hubei jiaoyu.

Chen Xionggen 陈雄根. 2000. "Guodian Chu jian Laozi Daqi mancheng shishi" 楚简老子大器曼成试释. *Zhongguo wenhua yanjiu suo xuebao* 中国文化研究所学报 9:237-44.

Chen Xiyong 陈锡勇. 2005. *Guodian Chu jian Laozi lunzheng* 郭店楚简老 子论证. Taipei: Taiwan Liren shuju.

Chen Zhu 陈柱. 1934. *Laozi* 老子. Shanghai: Shangwu.

_____. 1939. *Laozi Han Shi shuo* 老子韩氏说. Shanghai: Shangwu.庄子

Collingwood, R. G. 1945. *The Idea of Nature*. Oxford: Clarendon Press.

_____ 柯林伍德. 1999. *Ziran de guannian* 自然的观念. Translated by Wu Guosheng 吴国盛. Beijing: Huaxia.

Cook, Scott Bradley. 2012. *The Bamboo Texts of Gudian: A Study and Complete Translation*. Ithaca: Cornell University Press.

Cui Renyi 崔仁义. 1998. *Xingmen Guodian Chujian Laozi yanjiu* 荆门郭店楚简老子研究. Beijing: Kexue chubanshe.

Dai Kalin 戴卡琳. 2000. *Jiedu Heguanzi: Cong lunbianxue de jiaodu* 解读鶡冠子: 从论辩学的角度. Liaoning: Liaoningjiaoyu.

Davies, Paul. 1983. *God and the New Physics. New York: Simon & Schuster.*

_____ 保罗·戴维斯. 1995. *Shangdi yu xin wuli xue* 上帝与新物理学. Changsha: Hunan kexue jishu.

Ding Sixin 丁四新. 2002. "Chujian *Taiyi shengshui* yanjiu: Qianlun dangqian *Taiyi shengshui* yanjiu de zongti piping" 楚简太一生水研究：兼论当前太一生水研究的总体批评. In *Chudi chutu jianbo wenxian sixiang yanjiu*, edited by Ding Sixin. Wuhan: Hubei jiaoyu.

______. 2005. "Chu jian *Hengxian* zhangju shiyi" 楚简恒先章句释义. In *Chudi jianbo sixiang yanjiu* 楚地简帛思想研究, edited by Ding Sixin, vol. 2. Wuhan: Hubei jiaoyu.

______. 2010. *Guodian Chu zhushu Laozi jiaozhu* 郭店楚竹书<老子>校注. Wuhan: Wuhan daxue.

Ding Wei 丁魏 et al., eds. Forthcoming. *Laozi dianji kao* 老子典籍考.

Ding Yuanzhi 丁原植. 2005. "*Hengxian* yu gudian zhexue de shiyuan wenti" 恒先与古典哲学的始源问题. Paper Presented at the International Conference on Excavated Manuscripts and Pre-Qin Thought 新出土文献 与先秦思想重构国际学术研讨会. Taipei: National Taiwan University.

Dong Guangbi 董光壁. 1991. *Dangdai xin daojia* 当代新道家. Beijing: Huaxia.

Dong Lianchi 董莲池. 2000. "*Laozi* Daqi wancheng ji Daqi wucheng shuo puzheng" 老子大器晚成即大器无成说补证. *Guji zhengli yanjiu xuekan* 古籍整理研究学刊 5.

Dong San 董珊. 2004. "Chu jian Hengxian xiangyi liqiao jieshi" 楚简恒先 祥宜利巧解释. *Jianbo yanjiu* 简帛研究 2004/11.

Du Weiming 杜维明. 2004. "Cunyou de daxuxing: Zhongguo ren de ziran guan" 存有的连续性：中国人的自然观. *Shijie zhexue* 世界哲学 2004/1.

Fang Yong 方勇. 2011. *Zhuangzi* 庄子. Beijing: Zhonghua.

Feng Youlan 冯友兰. 2000. *Zhongguo zhexue xinbian shikao* 中国哲学史 新编试稿. Zhengzhou: Henan renmin,

Forke, Alfred.* 1972. *Lun-Heng: Wang Ch'ung's Essays.* New York: Paragon.

Fudan. 复旦大学出土文献. 2008. *Shangbo qi Fanwu liuxing zhongbian shiwen* 上博七凡物流形重编释文. Shanghai: Fudan daxue.

Gao Daowen 高道蕴. 1994. *Zhongguo caoqi de zhi sixiang* 中国早期的法治思想. In *Meiguo xuezhe lun Zhongguo falü zhuantong* 美国学者论中国 法律传统, edited by Gao Daowen et al. Beijing: Zhongguo zhengzhi daxue.

Gao Heng 高亨. 1988. *Laozi zhenggu* 老子正诂. Beijing: Zhongguo shudian.

______. 1989. *Laozi zhushi* 老子注译. Zhengzhou: Henan renmin.

Gao Ming 高明. 1996. *Boshu Laozi jiaozhu* 帛书老子校注. Beijing: Zhong-hua.

Graham, A. C.* 1960. *The Book of Lieh-tzu.* London: A. Murray.

______. 1986. *Yin-Yang and the Nature of Correlative Thinking.* Singapore: The Institute for East Asian Philosophies.

______. 1989. *Disputers of the Tao: Philosophical Argument in Ancient China.* La Salle, Ill.: Open Court Publishing Company.

_____ 葛瑞汉. 1998. *Yinyang yu guanlian siwei de benzhi* 阴阳与关联 思维的本质. Nanjing: Jiangsu guji.

_____. 2003. *Lundao zhe* 论道者. Translated by Zhang Haian 张海晏. Beijing: Zhongguo shehui kexue chubanshe.

Gu Shikao 顾史考. 2009. "Shangbo qi Fanwu liuxing jianxu ji yundu xiaopu"上博七凡物流形简序及韵读小补. www.bsm.org.cn/show_article.php? id=994.

Guo Yi 郭沂. 1998. "Cong Guidian Chu jian Laozi kan Laozi qiren qishu 从郭店楚简老子看老子其人其书. *Zhexue yanjiu* 哲学研究 7.

_____. 2001. *Guodian zhujian yu xianqin xueshu sixiang* 郭店竹简与先秦 学术思想. Shanghai: Shanghai jiaoyu.

Han Wei 韩巍. 2010. "Beida zang xihan zhushu ben Laozi de wenxian xue jiazhi" 北大藏西汉竹书本老子的文献学价值. *Zhongguo zhexue shi* 中国哲学史 2010/4.

_____. 2011. "Beida hanjian laozi jianjie" 北大汉简老子简介. *Wenwu* 文物 2011/6.

_____. 2012. *Beijing daxue zang xihan zhushu, er* 北京大学藏西汉竹, 书, 贰. Shanghai: Shanghai guji.

Hawkes, David.* 1959. *Ch'u Tz'u: The Songs of the South*. Oxford: Clarendon Press.

Hayek, Friedrich August von. 1975. *Kinds of Order in Society*. Menlo Park, Calif.: Institute for Humane Studies.

_____哈耶克. 1998. "Zisheng zifa zhixu yu wenming" 自生自发秩序与文明. Translated by Zheng Zhenglai 邓正来译. In *Ziyou yu zhixu: Hayek she hui lilun yanjiu* 自由与秩序——哈耶克社会理论的研究, edited by Deng Zhenglai 邓正来, Nanchang: Jiangxi Jiaoyu, p236—295.

Hegel, Georg Wilhelm Friedrich. 1892. *The Logic of Hegel*. Translated by William Wallace. Oxford: Clarendon Press.

_____. 1955. *Lectures on the History of Philosophy*. New York: Humanities Press.

_____ 黑格尔. 1981. *Xiao luoqi* 小逻辑. Beijing: Shangwu.

_____. 1983. *Zhexue shi jiangyan lu* 哲学史讲演录. Beijing: Shangwu.

Henricks, Robert.* 2000. *Lao Tzu's Tao Te Ching: A Translation of the Startling New Documents Found at Guodian*. New York: Columbia University Press.

Hightower, James R.* 1952. *Han Shih Wai Chuan: Han Ying's Illustrations of the Didactic Application of the Classic of Songs*. Cambridge, Mass.: Harvard University Press.

Hu Fenna 胡芬娜. 2002. "Guodian Chu mu zhujian *Laozi* Daqi mancheng shidu xianyi" 郭店楚墓竹简老子大器曼成释读献疑. *Yuyan xuekan* 语言 学刊 5.

Hu Shi 胡适. 1991. *Hu Shi xueshu wenji: Zhongguo zhexue shi* 胡适学术文 集: 中国哲学史. Beijing: Zhonghua.

______. 1998. *Zhongguo zhuantong zhong de ziran fa* 中国传统中的自然法. In *Zhongguo de wenyi fuxing* 中国的文艺复兴, edited bu Zou Xiaozhan 邹小站 et al. Changsha: Hunan renmin.

Huang Ruiyun 黄瑞云. 1995. *Laozi benyuan* 老子本原. Beijing: Renmin.

Ikeda Tomohisa 池田知久. 1993. "Chūgoku shisō ni okeru shizen no tanjō" 中国思想史における自然の誕生. *Zhongguo shehui yu wenhua* 中国 社会与文化 8.

______. 2006a. "Laozi de daoqi lun: Qi yu Mawangdui hanmu boshu ben" 老子的道器论: 基于马王堆汉墓帛书本. In *Chitian Zhijiu jianbo yanjiu lunji* 池田知久简帛研究论集, translated by Zao Feng 曹峰. Beijing: Zhonghua.

______. 2006b. *Mawangdui chutu wenxian shizhu congshu: Laozi* 马王堆 出土文献译注丛书•老子. Tokyo: Tōhō shoten.

______. 2009. *Daojia sixiang de xin yanjiu: Yi Zhuangzi wei zhongxin* 道家 思想的新研究: 以庄子为中心. Translated by Wang Qiyou 王启发 and Zao Feng 曹峰. Zhengzhou: Zhongzhou guji.

______. 2011. *Kakuten chikukan Rōshi no shin kenkyū* 郭店楚簡老子の 新研究. Tokyo: Kiko shoten.

Jaspers, Karl. 1953. *The Origin and Goal of History*. New Haven, Conn.: Yale University Press.

______ 雅斯贝斯. 1989. *Lishi de qiyuan yu mupiao* 历史的起源与目标. Beijing: Huaxia.

Jiang Rui 蒋瑞. 2000. "Shuo Guodian jianben Laozi Daoqi mancheng" 说郭店简本老子大器曼成. *Zhongguo zhexue shi* 中国哲学史 2000/1.

Jiang Xichang 蒋锡昌. 1988. *Laozi jiaogu* 老子校诂. Chengdu: Chengdu guji.

Knoblock, John, and Jeffrey Riegel.* 2000. *The Annals of Lü Buwei*. Stanford: Stanford University Press.

Legge, James.* 1962. *The Sacred Books of China: The Texts of Taoism*. New York: Dover.

Li Ling 李零. 2000. "Taiyi chongbai de kaogu yanjiu" 太一崇拜的考古研究. In *Zhongguo fangshu xukao* 中国方术续考, edited by Li Ling. Beijing: Dongfang.

______. 2002. *Guodian Chujian jiaodu ji* 郭店楚简校读记. Beijing: Beijing daxue.

______. 2003. "*Hengxian*" 恒先. In *Shanghai bowuguan cang Zhanguo Chu zhushu* 上海博物馆藏战国楚竹书, edited by Ma Chengyuan 马承源, vol. 3. Shanghai: Shanghai renmin.

Li Rui 李锐. 2004. "Qi shi zisheng: *Hengxian* dutede yuzhou lun" 气是自生: 恒先独特的宇宙论. *Zhongguo zhexue shi* 中国哲学史 2004/3.

_____. 2008. "Fanwu liuxing shiwen xinbian" 凡物流形释文新编. www.confucius2000.com/qhjb/fwlx1.htm.

Li Xiaoyu 李晓宇. 2003. "Guodian Chujian Taiyi shengshui puque yize" 郭店楚简太一生水补缺一则. *Sichuan daxue xuebao* 四川大学学报 2003/5.

Li Xueqin 李学勤. 1998. "Xingmen Guodian Chu jian suojian Guan Yin yishuo" 荆门郭店楚简所见关尹遗说. *Zhongguo wenwu bao* 中国文物报 4/29.

_____. 1999. *Taiyi shengshui de shushu jieshi* 太一生水的术数解释. In *Daojia wenhua yanjiu* 道家文化研究, vol. 17. Beijing: Sanlian shushe.

_____. 2003. "Chu jian *Hengxian* shouzhang shiyi" 楚简恒先首章释义. *Zhongguo zhexue shi* 中国哲学史 2003/3.

_____. 2004. "Chu jian *Hengxian* shouzhang shiyi" 楚简恒先首章释义. *Zhongguo zhexue shi* 中国哲学史 2004/3.

_____. 2006. *Kong Meng zhi yu Lao Zhuang zhi jian* 孔孟之间与老庄之间. Zhengzhou: Zhengzhou guji.

Liang Qixiong 梁启雄. 1985. *Hanfeizi qianjie* 韩非子浅解. Beijing: Zhong-hua.

Liang Qizhao 梁启超. 1999. "Laozi zhexue" 老子哲学. In *Liang Qizhao quanji* 梁启超全集. Beijing: Beijing chubanshe.

Liao Mingchun 廖名春. 2003. *Guodian zhujian Laozi jiaoshi* 郭店竹简 老子校释. Beijing: Qinghua daxue.

_____. 2004a. "Shangbo cang Chu zhushu *Hengxian* xinshi" 上博藏楚竹书 恒先新释. *Zhongguo zhexue shi* 中国哲学史 2004/3.

_____. 2004b. *Chutu jianbo congkao* 出土简帛丛考. Wuhan: Hubei jiaoyu.

_____. 2008. "Fanwu liuxing jiaodu lingli, yi" 凡物流形校读零札. www.confucius2000.com/qhjb/fwlx3.htm

Liao, W. K.* 1959. *The Complete Works of Han Fei Tzu*. London: Arthur Probsthain.

Lin Yizheng 林义正. 2005. "Lun *Hengxian* de yuzhou siwei: Ji yu neiguan gongfu de lingyi ge quanshi" 论恒先的宇宙思维:基于内观功夫的另一 个诠释. Paper Presented at the International Conference on Excavated Manuscripts and Pre-Qin Thought 新出土文献与先秦思想重构国际 学术研讨会. Taipei: National Taiwan University.

Lingxu Mingguang 另余明光. 1989. *Huangdi sijing yu Huang-Lao sixiang* 黄帝四经与黄老思想. Harbin: Heilongjiang renmin.

Liu Xiaogan 刘笑敢. 2006. *Laozi gujin: Wuzhong duijin yu xiping yinlun* 老子古今：五种对勘与析评引论. Beijing: Zhongguo shehui kexue yuan.

Liu Xinfang 刘信芳. 1999. *Xingmen Guodian Chudian Laozi jiegu*. Taipei: Yiwen yinshuguan.

Liu Zhao 刘钊. *Guodian Chu jian jiaoshi* 郭店楚简校释. Fuzhou: Fujian renmin.

Lou Yulie 楼宇烈. 1980. *Wang Bi ji jiaoshi* 王弼集校释. Vol. 1. Beijing: Zhonghua.

______. 1987. *Laozi Daode jing zhu* 老子道德经注. Beijing: Zhonghua.

Lovejoy, Arthur O. 1936. *The Great Chain of Being: The Study of the History of an Idea*. Cambridge, Mass.: Harvard University Press.

______ 拉夫乔依. 1996. *Ziran yixie hanyi* 自然的一些涵义. Translated by Peng Gang 彭刚. Beijing: Zhongguo shehui kexue.

Lu Jiaosan 卢育三. 1987. *Laozi shiyi* 老子释义. Tianjin: Tianjin guji.

Lu Yongpin 陆永品. 2008. "Laozi Daqi wancheng bianzheng" 老子大器晚成 辨证. *Zhongguo shehui kexue yuan yuanbao* 中国社会科学院院报 2008/4.

Luo Anxian 罗安宪. 2005. *Xujing yu xiaoyao: Daojia xinxing lun yanjiu* 虚静与逍遥: 道家心性论研究. Beijing: Renmin.

Lynn, Richard John.* 1999. *Classic of the Way and Virtue: A New Translation of the Tao-Te Ching of Laozi as Interpreted by Wang Bi*. New York: Columbia University Press.

Ma Chengyuan 马承源. 2001a. "Zhanguo Chu jian zhushu de faxian baohu he zhengli" 战国楚竹书的发现保护和整理. In *Shanghai bowuguan zang Zhanguo Chu jianshu, yi* 上海博物馆藏战国楚竹书(一). Shanghai: Shanghai guji.

Ma Chengyuan 马承源. 2001b. *Qianyan: Zhanguo Chu zhushu de faxian baohu he zhengli* 前言：战国楚竹书的发现保护和整理. Shanghai: Shanghai guji.

Ma Chengyuan 马承源, ed. 2008. *Shanghai bowuguan cang Zhanguo Chu zhushu* 上海博物馆藏战国楚竹书. Shanghai: Shanghai renmin.

Major, John S., Sarah A. Queen, Andrew S. Meyer, and Harold D. Roth.* 2010. *The Huainanzi: A Guide to the Theory and Practice of Government in Early Han China*. New York: Columbia University Press.

Mawangdui 马王堆汉墓帛书整理小组编. 1974. *Mawangdui Hanmu boshu, yi* 马王堆汉墓帛书(壹). Taipei: Wenwu.

______. 1976a. *Jingfa* 经法. Beijing: Wenwu.

______ 1976b. *Laozi* 老子. Beijing: Wenwu.

Nakajima Ryūzō 中嶋隆藏. 2005."Guanyu xianqin shidai heng de sixiang" 关于先秦時代恒的思想. Paper Presented at the Symposium on Confucian Studies 儒学全球论壇 Jinan: Shandong University.

Ōta Seiken 太田晴轩. 1972. *Laozi quanjie* 老子全解. In *Wuqiubeizhai Laozi jicheng xubian*, edited by Yan Lingfeng 严灵峰. Taipei: Yiwen yinshuguan.

Pang Pu 庞朴. 2004. "*Hengxian* shidu" 恒先试读. www.bamboosilk.org/ showarticle.asp?articleid=909.

Peng Hao 彭浩. 2000. "Yizhong xinde youzhou shengcheng lilun: Du *Taiyi shengshui*" 一种新的宇宙生成理论:读太一生水. In *Guodian Chujian guoji xueshu yantao hui lunwen ji* 郭店楚简国际学术研讨会论文集. Wuhan: Hubei renmin.

_____. 2001. *Guodian Chu jian Laozi jiaodu* 郭店楚简<老子>校读. Wuhan: Hubei renmin.

Plotinus. 1989. *On Nature, Contemplation, and the One*. Translated by Thomas Tyler. Edmonds, Wash.: Alexandrian Press.

_____ 普罗提诺. 2004. *Lun ziran, yisi he taiyi : Jiuzhang ji xuanshi ben* 论自然、凝思和太一：九章集选译本，石敏敏译. Beijing: Zhongguo shehui yanjiu yuan.

Qian Baochen 钱宝琛. 1932. "Taiyi kao" 太一考. *Yanjing xuebao* 燕京学报 1932/12.

Qian Yuzhi 钱玉趾. 2004. "Daqi wancheng, Daqi miancheng, Daqi wan-cheng" 大器晚成·大器免成·大器曼成. *Wenshi zashi* 文史杂志 5.

Qin Hualin 秦桦林. 2009. " Fanwu liuxing di ershiyi jian shijie" 凡物流形第二十一简试解.复旦大学出土文献与古文字研究中心. www.gwz.fudan. edu.cn.

Qiu Xigui 裘锡圭. 1995. *Guodian Chu mu zhujian* 郭店楚暮竹简. Beijing: Wenwu.

_____. 1996. "Mawangdui Laozi jiayi benquan qiangu shishu yu Daofajia: Qian lun Xinshu shang baixin wei Shen Dao Tian Pian zuopin"马王堆老子甲乙本卷前古佚书与道法家: 兼论心术上白心为慎到田骈作品. In *Shanggu sixiang, minsu yu guwenzi xueshi* 上古思想、民俗与古文字学史, by Qiu Xigui, 59-77. Shanghai: Shanghai Yuandong.

_____. 2004. "*Taiyi shengshui* 'Mingzi' zhang jieshi: Jianlun *Taiyi shengshyui* de fen-zhang wenti" 太一生水名字章解释:简论太一生水的分章问题. In *Zhongguo chutu wenxian shijiang* 中国出土文献十讲, edited by Qiu Xigui. Shanghai: Fudan daxue.

_____. 2008. "Shi *Hengxian* haishi *Jixian*?" 是恒先还是極先 *Proceedings of the 2007 International Symposium on Chinese Bamboo Manuscripts* 2007 中国简帛学国际论坛, 1-16. Taipei: Taiwan National University.

Rao Zongyi 饶宗颐. 1993. "Boshu Xici zhuan 'Daheng' shuo" 帛书系辞传大恒说. In *Daojia wenhua yanjiu* 道家文化研究, by Rao Zongyi, vol. 3. Shanghai: Shanghai guji.

Ren Jiyu 任继愈. 2006. *Laozi shidu* 老子绎读. Beijing: Beijing tushuguan.

Rickett, W. Allyn.* 1985. *Kuan-tzu: Political, Economic, and Philosophical Essays from China*. Vol. 1. Princeton: Princeton University Press.

_____. 1998. *Kuan-tzu: Political, Economic, and Philosophical Essays from China*. Vol. 2. Princeton: Princeton University Press.

Russell, Bertrand. 1966. *The Problem of China*. London: Allan & Unwin.

______ 罗素. 1996. *Zhongguo wenti* 中国问题. Shanghai: Xuelin.

Shi Huaci 史华慈. 2004. *Gudai Zhongguo de sixiang shijie* 古代中国的思想世界. Nanjing: Jiangsu renmin.

Strauss, Leo. 1953. *Natural Right and History*. Chicago: University of Chicago Press.

______列奥·施特劳斯. 2003. *Ziran quanli yu lishi* 自然权利与历史. Beijing: Sanlian shudian.

Sung, Z. D.* 1971. *The Text of Yi King*. Taipei: Chengwen.

Takeda Kenji 竹田健二. 2005. "*Kōsen* ni okeru ki no shisō" 恒先における 气の思想. In *Shikukan ga kataru kodai Chūgoku shisō* 竹简が语る古代中国思想, edited by Asano Yuichi. Tokyo: Kiko shoin.

Tan Baogang 谭宝刚. 2007. "Jin shinian lai guonei Guodian Chujian *Taiyi shengshui* yanjiu shuping" 近十年来国内郭店楚简太一生水研究述评. *Shixue yuekan* 史学月刊 2007/7,

Tan Jiefu 谭戒甫. 1964. *Mobian fawei* 墨辩发微. Beijing: Zhonghua.

Tang Lan 唐兰. 1975. "Mawangdui chutu Laozi yiben quanqian gushishu de yanjiu" 马王堆出土老子乙本卷前古佚书的研究. *Kaogu xuebao* 考古学报 1975/1.

Tong Shuye 童书业. 1982. *Xian Qin qizi sixiang yanjiu* 先秦七子思想研究. Jinan: Jilu shushe.

Wang Bo 王博. 1995. "Laozi ziran guannian de chubu yanjiu" 老子自然观念的初步研究. *Zhongguo zhexue shi* 中国哲学史 1995/3-4.

Wang Deyou 王德有, ed. 1994. *Laozi zhigui* 老子指归. Beijing: Zhonghua.

Wang Shumin 王叔岷. 2007. *Xian Qin daofa sixian jianggao* 先秦道法思想讲稿. Beijing: Zhonghua.

Wang Xianqian 王先慎. 1998. *Hanfeizi jijie* 韩非子集解. Beijing: Zhonghua.

Wang Zhongjiang 王中江. 2001. *Daojia xingershang xue* 道家形而上学. Shanghai: Shanghai wenhua.

______. 2002. "Laozi zhidao lishi laiyuan de yige tansao: Yi chuigong zhishi yu wuwei erzhi de guanlian wei zhongxin" 老子治道历史来源的一个探寻: 以垂拱之治与无为而治的关联为中心. *Zhongguo zhexue shi* 中国哲学史 2002/3.

______. 2006. *Zhongguo zhexue de zhuanhua yu fanshi* 中国哲学的转化与范式. Zhengzhou: Zhongzhou guji.

______. 2008. "Henxian de yuzhou guan ji renjian guan de gouzao" 恒先的宇宙观及人间观的构造. *Wenshi zhe* 文史哲 2:45-56.

______. 2009. "Fanwu liuxing bianlian xinzhi."凡物流形编联新见. www.jianbo.org/admin3/2009/wangzhongjiang001.htm.

_____. 1999. "Guodian Zhujian Laozi lveshuo" 郭店竹简老子略说. In *Zhongguo zhexue*(20) 中国哲学(第 20 辑). Shenyang: Liaoning Jiaoyu.

____. 2010a. "Dao yu shiwu de ziran: Laozi 'dao fa ziran' shiyi kaolun" 道与事物的自然：老子"道法自然"实义考论. *Zhexue yanjiu* 哲学研究 2010/8.

_____. 2010b. "Fanwu liuxing de guijun, guixin he guiyi"凡物流形>的"贵君"、"贵心"和"贵一. *Qinghua daxue xuebao* 清华大学学报 2010/1.

_____. 2012. *Taiyi shengshui* de yuzhou shengcheng moshi he tiandao guan" 太一生水的宇宙生成模式和天道观. In *Jianbo wenming yu gudai sixiang shijie* 简帛文明与古代思想世界, edited by Wang Zhongjiang, 31-55. Beijing: Beijing daxue.

Watson, Burton.* 1968. *The Complete Works of Chuang-tzu*. New York: Columbia University Press.

Wei Qipeng 魏启鹏. 2004. *Mawangdui Hanmu boshu Huangdi shu qianzheng* 马王堆汉墓帛书黄帝书笺证. Beijing: Zhonghua.

Wells, Marnix.* 2013. *The Pheasant Cap Master and the End of History: Linking Religion to Philosophy in Ancient China*. St. Petersburg, Fla.: Three Pines Press.

Wilhelm, Richard.* 1950. *The I Ching or Book of Changes*. Translated by Cary F. Baynes. Princeton: Princeton University Press, Bollingen Series XIX.

Wu Baihui 巫白慧. 2000. *Indu zhexue: Feituo jing tanyi he aoyi shu jiexi* 印度哲学:吠陀经探义和奥义书解析 . Beijing: Dongfang.

Wu Guang 吴光. 1985. *Huang-Lao zhi xue tonglun* 黄老之学通论. Hangzhou: Zhejiang renmin.

Xiang Zongzeng 向宗鲁. 1987. *Shuoyuan jiaozheng* 说苑校证. Beijing: Zhonghua shuju.

Xing Wen 邢文, ed. 2005. *Guodian Laozi yu Taiyi shengshui* 郭店老子与太一生水. Beijing: Xueyuan.

Xu Fuguan 徐复观. 2001. Zhongguo renxing lunshi: Xian Qin pian 中国人性论史:先秦篇. Shanghai: Sanlian shudian.

Xu Kangsheng 许抗生. 1985. *Bozi Laozi zhushi yu yanjiu* 帛子老子 注译与研究. Hangzhou: Zhejiang renmin.

Xu Zhuoyuan.许倬云. 2006. *Zhonguo gudai shehui shilun: Chunqiu Zhanguo shiqi de shehui liudong* 中国古代社会史论: 春秋战国时期的社会流. Guilin: Guangxi shifan daxue.

Yan Fu 严复. 1986. *Laozi pingyu* 老子评语. In *Yanfu ji* 严复集, vol. 4. edited by Wangshi 王栻. Beijing: Zhonghua shuju.

Yan Lingfeng 严灵峰. 1983. *Mawangdui boshu Laozi shitan*" 马王堆帛书老子试探. In *Wuqiu beizhai xuanji: Jingzi congzhu* 无求备斋选集·经子丛著, vol. 1. Taipei: Taiwan guoli pianshi guan.

Yanaka Shin'ichi 谷中信一. 2007. "Kōsen uchūron sekigi" 恒先宇宙論析義. *Chūgoku shutsutu zairyō kenkyū* 中國出土資料研究 11: 46-70.

______. 2010. "Shanghaibaku chikukan Hanbo ryukō kahon shakuchu" 上海博楚簡凡物流形甲本釋注. *Shutsutu zairyō to Kanji bunka kenkyū* 出土資料と漢字文化研究 5.

Yang Zesheng 杨泽生. 2009. "Fanwu liuxing zong shao de liangge zi"凡物流形从少的两个字. www.bsm.org.cn/show_article.php?id=999.

Yates, Robin D. S.* 1997. *Five Lost Classics: Tao, Huang-Lao, and Yin-Yang in Han China.* New York: Ballentine Books.

Yu Jiaokang 余敦康. 2005. *Zongjiao zhexue lunli* 宗教·哲学·伦理. Beijing: Zhonguo shehui kexue chubanshe.

Zhang Dainian 张岱年. 1982. *Zhongguo zhexue gang* 中国哲学大纲. Beijing: Zhongguo shehui kexue.

______. 1989. *Zhongguo gudian zhexue gainian fanchou yaolun* 中国古典哲学概念范畴要论. Beijing: Zhongguo shehui kexue yuan.

______. 2002. *Key Concepts in Chinese Philosophy.* New Haven, Conn.: Yale University Press.

Zhang Shunhui 张舜徽. 1982. *Zhou Qin daolun fawei* 周秦道论发微. Beijing: Zhonghua.

Zhao Jiangong 赵建功. 2005. "*Hengxian* yijie" 恒先易解. www.confucius2000.com/admin/list.asp?id=1541.

Zhao Jianwei 赵建伟. 1999. "Guodian Chumu chujian *Taiyi shengshui* shu-zheng" 郭店楚墓竹简太一生水疏证. *Daojia wenhua yanjiu* 道家文化研究 17.

Zheng Zhenglai 邓正来. 1998. Ziyou yu zhixu: Hayeka shehui lilun de yanjiu 自由与秩序：哈耶克社会理论的研究. Nanchang: Jiangxi jiaoxu.

Zhu Fengyu 朱凤翰. 2011. "Beijing daxue zang xihan zhushu gaishuo" 北京 大学藏西汉竹书概说. *Wenwu* 文物 2011/6.

Zhu Yuanqing 朱渊清, and Liao Mingchun 廖名春, eds. 2002. *Shangbo guanzang Zhanguo Chu zhushu yanjiu* 上博馆藏战国楚竹书研究. Shanghai: Shanghai shudian.

Zhu Yuanqing 朱渊清. 2002. "Ma Chengyuan xiansheng tan Shangbo jian" 马承源先生谈上博简. In *Shangboguan zang Zhanguo Chu jian zhushu yanjiu* 上博馆藏战国楚竹书研究, edited by Zhu Yuanqing and Liao Mingchun 廖名春. Shanghai: Shanghai shudian.

Index